SYMMETRY

and

THE END OF PROBABILITY

DangSon

Second edition 2003

Title: "Symmetry and the End of Probability"
Author: DangSon
Science – Philosophy

Created in 2000-2003 by DangSon

Second edition 2003

ISBN: 0-9740317-0-4
Library of Congress Control Number: 2003091876

Last Science Publishing
San Jose, California 95123
Contact: lastscience@yahoo.com
http://www.lastscience.com

PRINTED IN THE UNITED STATES OF AMERICA
by TD Printing, San Jose, California 95111
Cover design by David Romero

*to those who believe
there is more to science
than mechanical principles*

Symmetry and the End of Probability
Contents

ACKNOWLEDGEMENTS

Special thanks to Roger Liu, my former co-worker in Asia, who gave me the idea of keeping my writing style as accessible to the public as possible. Incredible as it may sound, it was this writing style that led to more than half of the ideas in this book. Needless to say, without Roger's advice these ideas may have never come to existence.

Jeanie Suzuki, my former boss and boss' boss in the semiconductor equipment industry, has been of great help in a different way. Discussions with her on seemingly unrelated subjects somehow changed the organization of this book altogether. Thanks to this radical change, the book became more complete.

Hanspeter Bleuler, my long-time friend and former colleague in the semiconductor equipment industry, deserves my gratitude for being the first serious believer in the Distribution theory after reading the first edition of this book. He gave me a surprise gift by completing a grueling computer simulation test without my asking. His test results are cited extensively in this edition.

Tim Starling, whom I do not know and may never meet, kept me focused with his strong and sometimes excited but very constructive criticism. The month-long daily email exchange between him (in Australia) and me (in the US) in September-October 2002 led to my decision to introduce the Principle of Symmetry in this edition. I had been reluctant to do this earlier, but in hindsight the addition of the Principle of Symmetry should improve the chance for the message in this book to reach both the public and the scientific community. The credit for this positive development naturally belongs to Tim.

Last but not least, I thank my friend from college days Nghi Huynh who provided me with an ideal environment to complete the rewriting of this book; and he did so wholeheartedly without even asking what I was trying to write. Minnesota winter is known to be very cold, but the winter of 2002-2003 will stay forever warm in my heart.

Foreword for the second edition

This book is a revision of the first part of the double-book "The End of Probability and the New Meaning of Quantum Physics", which was published through a print-on-demand house in August 2002 and is still available on the worldwide web at the time of this writing.

I was about to write a long foreword for this edition but, after careful consideration, I decided not to. It suffices to say that I have learned a few things from the first edition. This edition is my renewed effort to make it easier for the reader to judge my ideas on their merits.

Having said that, I do anticipate negative reactions, not only from diehard proponents of the Probability theory but also from many mainstream scientists. At the same time, I am very firm in my conviction that the ideas contained in this book will be a part of the next scientific revolution. It may take ten or twenty years, but that revolution will eventually come, because it has long been overdue.

The Probability theory has been a very useful tool of science for three centuries. I am personally grateful to the great men and women who have developed this branch of knowledge to its present state. Although this book calls for the end of probability so that science can move forward on a more solid footing, it could not have come to existence without the knowledge accumulated by the Probability theory.

As for the rest, I'll let the book speak for itself and I invite you, the reader, to judge its value.

Minnesota, December 2002
DangSon Tran

Chapter 1

A critical review
of the Probability theory

In this chapter, we will describe the extremely important role of the Probability theory in our lives and in science. We will review the four major interpretations of probability, namely subjective probability, classical probability, frequency probability, and the propensity interpretation of probability.

We will point out that all major theories of probability -except for subjective probability- suffer from the long-run paradox (long runs are theoretically possible but not seen in real life.) This paradox is an indication that something fundamental may be missing in the probability theories.

A. PROBABILITY AS A FACT OF MODERN LIFE

Open the sports page in the daily newspaper we see odds for football teams to win and lose, for player A to capture the next US open, for Brazil to win the next World Cup. Open the political page we see senator B is likely to beat his next opponent by 48%-44%, we see a slim chance for peace in the Middle East in the short term. Open the business page, we see predictions that the economy most probably will dip into a recession. Open a physics textbook, we find probability is the backbone of quantum mechanics, the modern branch of physics that brought us the transistor, semiconductor, computer, and this incredible information age. Turn our attention to genetics, believed to be the science of the 21st century, we again find probability as the integral part of its foundation.

No doubt about it, we live in The Age of Probability. We have taken probability for granted to the point that we no longer are aware of the fact that, as of the time of this writing, the Probability theory itself is still a pseudo-science. Probability is popular not because it is theoretically sound, but because it is successful, and success is hard to beat even in science. The popularity of Probability has caused serious concerns among scientists, including the most dominant scientist of recent time, Albert Einstein himself.

Since its conception the probability theory has always been under constant attacks by those who demanded mathematical rigor. To many probability is ignorance disguised as science. One of the most furious critics of the theory of Probability was the mathematician-philosopher and Nobel laureate Bertrand Russell in his book "Religion and Science".

History recorded that Einstein was shocked when his colleagues decided to use probability as the basis for their interpretation of quantum mechanics. His famous remark to Max Born in 1926 "I am convinced that God does not play dice" was an indirect expression of the low opinion he had for the Probability theory.

Until the so-called "Probability problem" is resolved, it is difficult for science to continue with the claim that it is the right path that will lead mankind to the ultimate understanding of the physical universe.

In this and the next chapter, we will take a step back to understand the problems plaguing the Probability theory. We will propose solutions to these problems in subsequent chapters.

B. THE PROCESS OF COIN TOSSING

To start our discussion we will choose the scientific problem of predicting the outcome of the next coin toss. Since this is the simplest problem that can be analyzed by the Probability theory in a meaningful way, it shortens our discussion with no risk of generality because the same logic can be generalized to all cases involving probabilities, including the hypothetical case of continuous probability functions[1].

C. PROBABILITY AND THE LONG RUN PARADOX

The failure of classical physics
in predicting coin tossing outcomes

There is no doubt that the motion of a tossed coin from the time it leaves the hand of the tosser to the time it settles on either heads or tails can be formally described by the laws of classical physics. More specifically, the motion of a coin must obey Newtonian mechanics. The problem is, the motion of a flipped coin is so complex that a slight variation in condition could change a tendency toward heads to a tendency toward tails, and vice versa.

It is tempting to suggest that a chaotic approach (i.e., setting up the equation of motion with very fine resolutions in space and/or time to take high non-linearity into account) is all it takes to solve this problem. This is not true because the flipped coin "digitizes" all effects, no matter how minute, into only two values "heads" and "tails"; and no matter how fine we resolve the space-time increment, there are always potential situations that require yet finer resolutions to yield correct predictions. Thus, in the general case we will end up with a non-zero set of tosses with undecided outcomes. Adding the difficulty in determining the initial conditions of the tossing process, which is usually not known ahead of time and could vary wildly from toss to toss, we must conclude that the chaotic approach is not a practical solution to the problem of predicting coin toss outcomes.

Since the chaotic approach is now considered a part of classical physics, it is no surprise that it was the failure of classical physics in predicting coin toss outcomes and many related problems that gave rise to the Probability theory about 300 years ago.

Probability conventions and Kolmogorov's axioms

The failure of classical physics forced scientists to look for an alternative method to account for processes such as coin tossing. This alternative method eventually evolved into what is now known as the Probability theory.

To cast the concept of probability into the language of mathematics the following conventions have been adopted:

Certainly true, certainly will happen, etc. means probability value is equal to 1.

Certainly not true, certainly will not happen, etc. means probability is equal to zero *(however, an event with probability=0 may happen, and an event with probability=1 may not happen[2].)*

Although there are many probability interpretations, the consensus is that the theory of probability can be built on three axioms, known as the Kolmogorov's axioms operating on elements x_1, x_2, x_3, etc. of an applicable set S:

1. $0 \leq P(x_i) \leq 1$
   The probability that x_i is true $P(x_i)$ must lie between 0 and 1.
2. $P(S) = 1$
   The probability that at least one element in S is true is 1.
3. $P(x_1$ or x_2 or x_3 ... or $x_n) = \Sigma P(x_i)$ if all x_i's are independent.
   If x_1, x_2, x_3, ..., x_n are independent, the probability that at least one of them is true is the sum of the individual probabilities.

The limited value of subjective probability

The concept of probability arises naturally from the reality of everyday's life, where it can be considered as a "reasonable person's" assessment of unpredictable future events. This popular form of probability is usually called "subjective probability" to differentiate it from other interpretations that will be introduced shortly.

In the case of coin tossing, depending on our degree of belief when we make a prediction, the probability for heads that we choose could be, say, 0.3 for the first toss and 0.9999 the second toss. Mathematically this means we lean toward "tails" for the first toss (because 0.3 is closer to 0, which represents tails, than 1, which represents heads;) and we strongly believe that the second toss will be heads (because 0.9999 is very close to 1.)

While the meaning of subjective probability seems to be closest to our understanding of the word "probability", this concept has practical drawbacks. First, as the name implies, if we ask any two persons to give their probability value for the same toss, it is almost certain that their predictions will be different from each other.

The second problem with subjective probability is best illustrated by means of an example. Let's say we predict "probably heads" for the next toss and the outcome is tails. Does this mean our prediction is wrong? No! Because we did not predict "certainly heads". From this example, it is clear that subjective probability predictions are never verifiable.

For these reasons and others, subjective probability has very limited application as a scientific concept, at least in the meaning of "scientific concept" that we know today.

The indifference principle and classical probability

Let's look at a series of random tosses performed with the same coin. If the construction of the coin looks regular, there would be no reason to assume that heads is preferred over tails or vice versa. Further, since there is no clear reason for the coin to remember past history, it seems reasonable to assume that the probability for heads or tails to turn up is the same for every toss.

This reasoning is the essence of the "indifference principle", which states that if there is no compelling reason to believe that a difference should exist among N alternatives, the probability of occurrence of each alternative must be taken to be the same as those of other alternatives. The "indifference principle" is taken for granted as a first principle of an interpretation of probability now known as "classical probability".

It can be seen that classical probability is an effort to improve over subjective probability -- by simply removing all subjectivity from the determination of probability values. Although classical probability is usually credited to Simon Laplace who popularized it in the early part of the 19th century, historically it went back at least to the year 1718 with Abraham de Moivre's "The Doctrine of Chance".

Classical probability and the equal opportunity principle

Once the indifference principle is accepted, the probability values of combined events can be calculated in a straightforward manner. For example, if we let P(x) stand for the probability that "x" takes place, then for two events a and b that appear to have nothing to do with each other, the following can be deduced from Kolmogorov's axioms:

$$P(a \text{ or } b) = P(a) + P(b) \tag{1}$$

$$P(a \text{ and } b) = P(a) \times P(b) \tag{2}$$

We will define an unbiased coin as a coin whose construction creates no clear preference for either heads or tails. By applying (1) to one toss of this coin with "a"=heads, "b"=tails, since the outcome must be either heads or tails, it is obvious that:

$$P(a \text{ or } b) = P(\text{heads or tails}) = 1 \tag{3}$$

Equation (1) becomes:

$$1 = P(\text{heads}) + P(\text{tails}) \tag{4}$$

Since the coin is unbiased, it follows that:

$$P(\text{heads}) = P(\text{tails}) = 0.5 \tag{5}$$

It is therefore expected that in a long series of tosses, the ratios for heads as well as tails will be approximately 0.5.

For two tosses at a time there are 4 possible combinations, with only one way to get two heads (HH) or two tails (TT), but two ways to get one heads one tails (HT, TH). The corresponding ratios are: Two heads $0.5 \times 0.5 = 0.25$, one heads one tails $0.5 \times 0.5 + 0.5 \times 0.5 = 0.50$, two tails $0.5 \times 0.5 = 0.25$.

For three tosses at a time there are 8 possible combinations, with one way to get three heads (HHH) or three tails (TTT), three ways to get 2 heads one tails (HHT, HTH, THH) or one head two tails (HTT, THT, TTH). The ratios then are: Three heads 1/8, two heads one tails 3/8, one heads two tails 3/8, three tails 1/8.

For four tosses at a time there are 16 possible combinations, with one way to get four heads (HHHH) or four tails (TTTT), four ways to get 3 heads 1 tails (HHHT, HHTH, HTHH, THHH) or 1 heads 3 tails (TTTH, TTHT, THTT, HTTT), six ways to get 2 heads 2 tails (HHTT, HTHT, HTTH, TTHH, THTH, THHT). The ratios then are: Four heads 1/16, three heads one tails 4/16, two heads two tails 6/16, one heads three tails 4/16, four tails 1/16.

In the example below, an unbiased "coin" was "tossed" by a computer (i.e., computer simulation of a coin toss.) The actual ratios are quite close to the ratios predicted by classical probability.

TABLE 1: Experimental and ideal convergence ratios**
(test date: August 2001)
40,000 trials of 1 toss each

Heads	Tails	Count	Ratio (actual)	Ratio (ideal)	Error*
1	0	20,278	0.507	0.500	1.4%
0	1	19,722	0.493	0.500	-1.4%
		40,000	**1.000**	**1.000**	

1,500 trials of 2 tosses each

Heads	Tails	Count	Ratio (actual)	Ratio (ideal)	Error*
2	0	357	0.238	0.250	-4.8%
1	1	753	0.502	0.500	0.4%
0	2	390	0.260	0.250	4.0%
		1,500	**1.000**	**1.000**	

2,600 trials of 3 tosses each

Heads	Tails	Count	Ratio (actual)	Ratio (ideal)	Error*
3	0	305	0.117	0.125	-6.2%
2	1	953	0.367	0.375	-2.3%
1	2	983	0.378	0.375	0.8%
0	3	359	0.138	0.125	10.5%
		2,600	**1.000**	**1.000**	

2,400 trials of 4 tosses each

Heads	Tails	Count	Ratio (actual)	Ratio (ideal)	Error*
4	0	143	0.0596	0.0625	-4.7%
3	1	615	0.2563	0.2500	2.5%
2	2	924	0.3850	0.3750	2.7%
1	3	560	0.2333	0.2500	-6.7%
0	4	158	0.0658	0.0625	5.3%
		2,400	**1.0000**	**1.0000**	

* Error defined as (Actual – Ideal)/ Ideal
**Readers who still have doubt may want to perform the same experiment themselves. While the exact counts of heads and tails are never repeatable, the ratios will compare well with their ideal values each time. They get closer if the numbers of trials are increased.

The fact that the calculated ratios are well approximated by the actual ratios explains why classical probability still plays a major role in science today despite its conceptual problems. It also appears to imply that Nature treats all distinct combinations with the same probability (e.g., HH, HT, TH, TT) equally in the long run. This we will call the "equal opportunity principle". While the logic behind the equal opportunity principle is best considered a mystery, it is clear that it must be strongly related to randomness. Not surprisingly, classical probability takes the position that the equal opportunity principle is due completely to randomness. We will question this position in a later chapter.

The long-run paradox of classical probability

Despite its successes, classical probability leads to a result that is not corroborated by reality. The reader may want to verify that, by applying (2) successively, the probability P(N/N) to get N heads by tossing an unbiased coin N times is:

$$P(N/N) = 0.5^N \qquad (6)$$

It is clear from (6) that, although P(N/N) decreases exponentially with N, it is never zero. For example, despite the minute value of P(100/100) = 0.5^{100}, it is possible for the next 100 tosses to be heads. Since N could be arbitrarily large, classical probability cannot rule out the possibility of, say, a million heads in a row in the next million tosses. The problem is, in reality the heads-to-total ratio has always converged to 0.5 rather quickly and no one has ever reported a case of say, 100 heads or tails in a row in actual coin tosses.

The writer himself has performed many simulated coin toss experiments on two window-based personal computers (with unbiased 1's and 0's in place of heads and tails). The longest strings produced by the two computers were 25 and 33 respectively.

Table 2 shows an independent test performed on a Macintosh computer and communicated to the writer in November 2002. In this test approximately 141.3 billion unbiased tosses were simulated and the strings of heads were combined in the finally tally. For short strings the probability predictions are well approximated by the actual counts, but significant deviations start to show up with strings longer than 20. Most seriously, strings longer than 28 are completely absent[3].

While strings longer than 28 are expected with more powerful random number generators, it is doubtful if a string of, say, 1000 will ever be seen in the near future.

TABLE 2: Computer simulation of 141.3 billion coin tosses
(by Hanspeter Bleuler, Germany, private communication, November 2002)
Counts are combined total for strings that are all-heads

STRING	PROBABILITY PREDICTION	ACTUAL COUNT
1	70,652,392,105	70,650,390,138
2	35,326,196,053	35,326,178,115
3	17,663,098,026	17,663,595,966
4	8,831,549,013	8,832,056,857
5	4,415,774,507	4,416,137,049
6	2,207,887,253	2,208,074,053
7	1,103,943,627	1,104,163,729
8	551,971,813	552,131,003
9	275,985,907	276,011,526
10	137,992,953	137,874,262
11	68,996,477	69,127,258

12	34,498,238	34,545,871
13	17,249,119	17,287,151
14	8,624,560	8,645,262
15	4,312,280	4,311,566
16	2,156,140	2,118,205
17	1,078,070	1,055,064
18	539,035	550,366
19	269,517	265,437
20	134,759	132,277
21	67,379	62,715
22	33,690	34,000
23	16,845	18,727
24	8,422	10,795
25	4,211	2,870
26	2,106	1,309
27	1,053	2,118
28	526	521
29	**263**	**0**
30	**132**	**0**
31	**66**	**0**
32	**33**	**0**
33	**16**	**0**
34	**8**	**0**
35	**4**	**0**
36	**2**	**0**
37	**1**	**0**

For reference purposes, we will call this discrepancy "the long-run paradox of classical probability" [4].

Proponents for classical probability have several explanations for the long-run paradox. In the next chapter we will see that these explanations are actually in conflict with the foundation of probability.

The incompleteness of frequency probability

The other major school of probability is the relative frequency interpretation of probability, which we will call "frequency probability" for brevity. Frequency probability became a major player in the first half of the 20th century thanks largely to the works of Richard von Mises[5]. Dissatisfied with classical probability, von Mises created two axioms for frequency probability:

1. Axiom of convergence[6] (or convergence axiom): The actual ratio of a chosen outcome will approach its ideal value if the sequence is long enough.

    *Example: The ideal ratio N_H/N for an unbiased coin will approach 0.5 with enough tosses.

2. Axiom of randomness (or randomness axiom): The outcomes must be randomly distributed among the trials.
   *Meaning: It is impossible to predict when a particular outcome will occur in a sequence.
   **Note: The axiom of randomness actually was referred to as "the Principle of Randomness" by Richard von Mises[5] in his original treatment of probability. Here we have chosen to use "axiom" for uniformity of terms. Von Mises also gave this axiom an alternative name "The Principle of Impossibility of a Gambling System"; because to von Mises, randomness means there exists no system that can successfully predict a future outcome.*

In frequency probability, probability is *defined* as the ratio of the *ideal* number of heads N_{H0} over the total number of trials N:

$$P(N_{H0}/N) = N_{H0}/N \qquad (7)$$

It can be seen that frequency probability was designed to fit empirical data. Specifically, frequency probability avoids the long-run paradox by invoking the convergence axiom. Since table 1 can also be explained in the framework of frequency probability, it appears that frequency probability is an improvement over classical probability. Unfortunately, despite von Mises' objection to the indifference principle, when it comes to prediction of long-term ratios, frequency probability ends up using the same procedure as classical probability[7]. This makes frequency probability no more than a bandage fix for classical probability.

Not surprisingly, frequency probability cannot identify any factor or factors that decide a "long enough sequence". It is still possible, for example, in the mathematical framework of frequency probability to have a long run of one million consecutive heads. Thus, in reality, frequency probability has swept the long-run paradox under the rug instead of solving it.

The incompleteness
of the propensity interpretation of probability

So far we have been led to believe that the outcome of each coin toss and the long-term result of many tosses are manifestation of some inherent property of the coin (e.g., its construction) and have nothing to do with the experimental condition. If this were true, there would be no difference between the tossing process performed by, say, a three years old child and that by an advanced robot arm which can be precisely programmed for specialized tasks. In reality, it is doubtful if a three years old child can cause the long-term heads-to-total ratio to deviate significantly from 0.5; whereas a modern robot arm, with proper programming, could very much give us whatever long-term ratio we desire, including heads all the time or tails all the time.

This kind of process-dependency seems to have been recognized by Karl Popper in his propensity interpretation of probability[8]. Unfortunately, the propensity interpretation only has value as a conceptual improvement over its predecessors [9, 10], and it still leaves the long-run paradox unsolved.

E. THE PROBABILITY THEORY TODAY

Strengths and weaknesses of major probability interpretations

Let's summarize what we have learned so far:

1. Subjective probability: Probability values assigned to a simple or complex event by personal criteria. Probability values tend to vary from case to case and could be very subjective. Also called "personal probability".

 * *Strength: Close to original meaning of the word "probability".*
 * *Weakness: Subjective and unverifiable, therefore difficult to apply to existing science.*

2. Classical probability: Fixed probability values assigned to simple or complex events according to the indifference principle. Probability values are verified by long-term results.

 Strength: Mathematically simple and can be applied to science.
 Weakness: Suffers from the long-run paradox

3. Frequency probability: In contrary to classical probability, frequency probability takes the position that the concept of probability does not have any meaning for a simple or complex event, and should be applied only to a series of events. It is believed that "long enough" series will converge to long-term probability values.

 * *Strength: Fits empirical results so far.*
 * *Weaknesses: Long-term convergence is justified by empirical result only, with no logical foundation. Still must rely on the method of classical probability to calculate probability values. Long-run paradox still exists because the theory cannot quantity the meaning of "long enough sequence".*

4. Propensity interpretation of probability: Probability reflects a tendency related to experimental condition.

 Strength: Includes role of experimental condition, which was unaccounted for in classical and frequency probability.
 Weaknesses: Lacks mathematical formalism[10]. Does not solve the long-run paradox.

A critical review of the probability theory

The confusing state of the Probability theory today

With the first three interpretations being quite popular, and the last gaining strength in texts on physical sciences such as quantum physics, the application of probability in life and in science today is at best confusing. In recent years a number of textbooks started using "odds" and "chances" for frequency probabilities. Unfortunately, due to the lack of an agreed standard, this practice only adds to the confusion.

A few examples of the confusing state of probability today:

Instead of saying "the probability for the next toss to turn up heads is 1/2", some say "the odds for the next toss to be heads is one-to-one", meaning that the odds for success and failure are equal. Although the later statement sounds like frequency probability, it is actually a subjective or classical probability statement since we mean to apply it to a single event (the next toss).

In everyday life, we use expressions such as "chances are, the train will be late", "the odds are against my home team in the next game". These are our judgments of future outcomes based on known past histories and present information. Note, however, that although these statements sound like frequency probabilities, they are our personal predictions for individual events and therefore are subjective probabilities.

When the statistical model predicts that 10% Asian male workers will be unemployed in an economic downturn (a frequency probability prediction), a typical statement that we hear on radio or TV would be "A young Asian American male will have a 1-in-10 chance to be unemployed." Since this statement assumes that all young Asian American males have the same probability of being unemployed regardless of their education and skill level, it is an (absurd) classical probability statement disguised in the language of frequency probability.

In quantum physics it is common to verify a hypothesis with combined results of a series of experiments, which is an application of frequency probability. However, physicists tend to make statements such as "the probability that an electron will emit light is approximately v/c," which are classical probabilities.

In this confusing state of affair, the only thing that we can be sure is that there is no universal description of probability in existence today. Although some users of the "Probability theory" believe that they are working with a unified theory, in reality they only patch together bits and pieces of the four interpretations that we have just presented.

Special note on subjective probability and the long-run paradox

Interestingly, there is no conflict between subjective probability and the long-run paradox. In fact, it is human nature to believe that many consecutive occurrences of, say, black on a roulette table "causes" the probability for red to increase steadily, and when the string of black reaches certain length, the probability for red will become certainty (hence no long-run paradox for black). This popular gambling strategy, which is an example of subjective probability, is often ridiculed by proponents of classical probability and frequency probability as "the gambler's fallacy". Since the long-run paradox raises the possibility that classical probability and frequency probability may be wrong, it also forces us to abandon their judgment and return "the gambler's fallacy" to the status of an unproven hypothesis.

We face a peculiar possibility. If the long-run paradox turns out to be a problem that cannot be resolved within the framework of existing probabilistic mathematics, the only probability interpretation that is left standing after our analysis may be the largely ignored and often ridiculed "subjective probability". The reader will see that this is indeed the case.

Conventions for further discussion

Because of its unique ability to deal with the long-run paradox, we will separate subjective probability from the other three probability interpretations in all of our discussions.

With this understanding, from here on forward when we say "the Probability theory" without qualification, the term should be understood as "the most logical synthesis of classical probability, frequency probability, and propensity". The term "probability" without qualification should also be understood as "probability as defined by the most logical synthesis of classical probability, frequency probability, and the propensity interpretation of probability".

F. THE NEXT STEP

It is clear from our analysis in this chapter that none of the existing probability interpretations can give a satisfactory account of the class of phenomena that it is supposed to describe. This problem has been largely ignored by many members of the scientific community who argue that we can always patch the various probability theories together to meet specific needs as they arise; and that is good enough for science.

These probability proponents have forgotten that the long-run paradox contradicts the foundation of the three probability interpretation of interest (i.e., classical , frequency, and Popperian propensity), and contradiction is not acceptable in science. Thus, the long run paradox is an indication that the very foundation of probability may be fundamentally flawed. We will examine this possibility in the next chapter.

First written Dec 2000
Revised July 2001, September 2001, November 2002
Revised for second edition November 21, 2003
©DangSon. All rights reserved

NOTES:

1. Our argument for two possibilities (e.g., heads and tails) can be generalized to multiple (but still discrete) possibilities in a straightforward manner. Continuous probability functions correspond to the case where the number of possibilities is expanded to infinity Although there is no change in basic concepts going from discrete to continuous probability; the later is most conveniently handled by the method of calculus.

2. We know that an event that cannot happen must have probability value zero, and an event that certainly will happen must have probability value 1. Curiously, the reverse is not true. This is a conceptual ambiguity caused by the point axiom of mathematics; according to which the number of points on any given line segment is infinity.

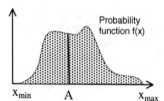

Probability function f(x)

x_{min} A x_{max}

Figure 1a: Event A definitely could happen, but in probability calculus the probability of its happening is zero. This is the inherent ambiguity built-in with probability calculus.

Here probabilities are presented as areas under a continuous probability function f(x) that covers all possibilities. Let's focus on an event that has just taken place. The fact that it has just happened confirms that it could happen. Now this event must correspond to a particular point A along the horizontal line; but it is clear that the area under the curve associated with point A is zero. Conversely, since A happened, the event "not A" did not happen, despite its probability value P(not A) = 1-P(A) = 1-0=1. We thus have the curious situation that an event that cannot happen has probability value zero, but an event with probability value zero could happen; likewise an event that certainly will happen has probability 1, but an event with probability 1 may not happen.

This ambiguity reminds us that mathematics is just a tool that we invented to describe Nature, and every tool has its limit. We must be aware of this limit to avoid mistakes when we apply mathematics to science.

In explaining why an event -such as 1000 consecutive heads- has never been seen, it is usually argued: "Because the probability for its occurrence is too low." This argument is so commonly used, even by probability experts, that no one seems to notice that it is logically invalid. Its invalidity can be established by *reductio ad absurdum* as follows: "If low probability is the reason why an event has not occurred for a long long time, then it should be forever before an event with zero probability to occur. Since zero probability events occur all the time, the logic is flawed."

3. The probability to get N heads or tails in a row is $P(N/N) = 0.5^N$. For N=29, in 141.3 billion tosses classical probability predicts 263 occurrences of strings of 29 heads. Since the actual count is zero, we must question whether classical probability is the correct description of the computer simulated coin tossing process. Proponents for classical probability will counter-argue that the computer is not a valid tool to test the Probability theory. We will show why this is not a sound argument in the next chapter.

4. We will show in the next chapter that the possibility of achieving an arbitrarily large number of consecutive heads is in conflict with the Law of Large Number, which asserts that when the number of tosses approaches infinity, the heads/tails ratio will approach its intended value (about 0.5 for an unbiased coin). This is a serious blow to classical probability as the Law of Large Number has always been assumed to be one of its results.

5. See, for example, "Probability, Statistics, and Truth", Richard von Mises, book, Dover 1981. This is an unabridged reproduction of the 1957 English translation of the 1951 German original; which in turn is a revision of the 1928 German original.
Von Mises refuted the "classical probability interpretation" of De Moivre and Laplace (i.e., all mathematically equivalent events have the same probability to occur) and championed the "relative frequency interpretation", referred to as "frequency probability" in this book. According to frequency probability the concept of probability should be applied only to series of events, not single events. Von Mises used the term "collective" to refer to a series of events.

6. The convergence axiom of frequency probability is clearly an *ad hoc* axiom because convergence is taken for granted by appealing to empiricism.

7. Strictly speaking, frequency probability must rely on actual experimental results to establish long-term ratios. But this would make it inferior to classical probability in predictive power in the case of coin tosses, for example. In practice frequency probability is usually complemented by classical probability in long-term distribution predictions.

8. See, for example, the article "Propensity, Probabilities, and the Quantum Theory", Karl Popper, 1957; reprinted in "Popper Selections", edited by Miller, Princeton University Press, 1985.

9. The following quote -from the document cited in 8 above- explains Popper's intention for the term "propensity".

A critical review of the probability theory

"We thus arrive at the propensity interpretation of probability. It differs from the purely statistical or frequency interpretation only in this –that it considers the probability as a characteristic property of the experimental arrangement rather than the property of the sequence.

"The main point of this change is that we now take as fundamental the probability of the result of a single experiment, with respect to its conditions, rather than the frequency of results in a sequence of experiments."

While the meaning of the second paragraph is somewhat vague, it appears that Popper meant to apply the concept of propensity uniformly to each individual event. This makes propensity functionally equivalent to classical probability with the added improvement that experimental conditions are taken into account.

The writer is puzzled by this choice, considering that in "The Logic of Scientific Discovery" (German original 1935, English translation Harper Torchbooks, 1965) Popper had been building on a modification of frequency probability by adding the concept of n-freedom. Why did Popper, in his propensity interpretation, did not take the view that propensity is the realization of an experimental condition, and this realization requires a long enough sequence of events? In other words, propensity is a frequency probability interpretation with strong emphasis on experimental condition. This would seem to be the most logical synthesis, which in the writer's opinion is also the most fitting model for, say, quantum mechanics.

It may sound unfair to guess someone's intention when he is not available to defend himself, but the writer suspects that Popper's notorious ego did not allow him to accept that his prized discovery "propensity" was only a modification of a larger discovery, i.e., frequency probability.

10. A Physicalist's Interpretation of Probability, Laszlo E. Szabo, talk presented at the Philosophy of Science Seminar, Eötvös, Budapest, 8 October 2001.

In this talk, Szabo made the following objection against the propensity interpretation of probability.

"In Propensity Interpretation, probability - propensity - is a separate quantity, which is not expressed in terms of other, empirically defined quantities. How, then, is the numerical value of propensity determined? We have no starting point for the empirical test of the value of propensity. Consequently, there is no empirical basis for such a proposition as ``the probability of getting <Heads> is 1/2", and the whole talk about probabilities loses empirical control. We do not even know whether propensities satisfy Kolmogorov axioms, or not."

Chapter 2

The paradoxes of probability

We start this chapter with a brief review of several important concepts in statistics: The differentiation between deterministic process and distributive process (a new name in place of the old name "random process"), binomial and normal distributions .

We will argue that event probability is a valid concept, but classical probability leads to paradoxes which remain unsolved in frequency probability and Popper's propensity interpretation of probability. In addition, we will show that two concepts believed to be logical consequences of the Probability theory, namely the Law of Large Number and the "Physically Impossible Rationale", are actually incompatible with probability.

Based on these results we will conclude that, as a scientific theory, probability is either wrong or incomplete. We will state our intention to search for a new theory that can solve the long-run paradox and is compatible with both the Law of Large Number and the "Physically Impossible Rationale".

A. IMPORTANT CONCEPTS IN STATISTICS

Deterministic process and distributive process

A distributive process is a process whose outcome is not single-valued (as in deterministic processes), but is multi-valued and can be thought of as a distribution of multiple outcomes. As the reader may have guessed correctly, these are the exact processes covered by the Probability theory (and often referred to as "random processes").

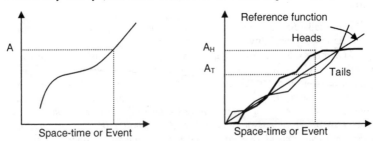

Figure 1: In a deterministic process, once space and/or time is specified we know what state the process is in, because there is a unique value for it (left picture). In a distributive process, however, there are multiple values; and the data form a distribution of many combinations (right picture).

The right graph of figure 1 illustrates the process of tossing a coin. Each point in time corresponds to a different distribution in cumulative heads and tails (A_H and A_T in graph). Both have to be taken together as a combined outcome. Thus, the process is distributive. In contrast, the process in the left graph is single-valued. It is a deterministic process, because at any given time there is only a single outcome A.

Distributive entity

By "distributive entity" we mean an entity whose behavior is not governed by deterministic laws. For example, a coin is a distributive entity (in a coin tossing experiment) because no deterministic law can predict without fail whether the next toss will be heads or tails.

Binomial distribution

We have touched on binomial distribution in the last chapter. Binomial distribution is the ideal distribution that we obtain when we toss N coins at a time. Recall that there are N+1 possible combinations with N coins. They are:

N heads and no tails
N-1 heads and 1 tails
N-2 heads and 2 tails
....
1 heads and N-1 tails
0 heads and N tails

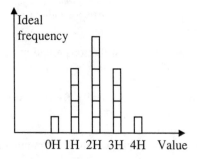

Figure 2: If we flip 4 coins at a time, we will get 5 possible combinations 0H4T, 1H3T, 2H2T, 3H1T, and 4H0T. We present these on the horizontal direction, and the relative abundance for them in the vertical direction. This ideal graph is called a distribution curve. In case of real data, we would call such a graph a histogram.

Depending on the biasedness of the coin (and the process, too, as we will see in a later chapter) each combination would have a different ideal ratio. For example if we toss 4 unbiased coins 4 times to get a total of 16 outcomes, the ideal distribution would be: One 4-heads (ideal ratio = 1/16), one 4-tails (1/16), four 3-head-1-tails (4/16), four 1-heads-3-tails (4/16), six 2-heads-2-tails (6/16). Since the world is not ideal, the actual distribution will deviate from the ideal distribution. We will return to this point later, but for now it is not our concern.

The ideal distribution for 4 unbiased coin is shown in figure 2. This is the standard way to present a distribution, ideal as well as real:

"Value" (such as 3 heads) on the horizontal axis, "count" (or frequency on the vertical axis). For example, from figure 2 the count for "value" 4H (four heads) is 1, which could stand for any unit we choose, say, 1.5, 100, 1000, one million, etc. By multiplying the chosen unit and the "value" we get the desired number for a combination.

Normal distribution

If a large number of coins are tossed in each trial, the vertical bars in the binomial distribution are pushed closer together. When the number of coins approaches infinity the bars becomes lines of zero thickness, giving us a continuous curve. This limiting case is called the normal distribution. The normal distribution is convenient for data analysis for three reasons:

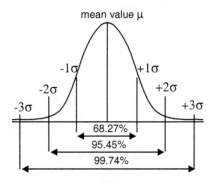

Figure 3: The great thing about the normal curved (also called the Gaussian curve) is that it is a "one size fits all" tool that requires us to know only two parameters: Mean and sigma. In addition, each number of sigma's corresponds to a fixed percentage of possibilities.

1. The normal distribution fits many statistic distributions found in real life. We will return to the significance of this point in a later chapter.

2. All normal distributions can be standardized with a fixed measure called sigma (σ) which is an average deviation of all data points from the mean value of the distribution.

3. The standardized normal distribution curve extends from minus to plus infinity. However, the range (-1σ, $+1\sigma$) already contains 68.3% of the total counts (visually seen as the area under the bell-shaped curve), the range (-2σ, $+2\sigma$) about 95.4%, the range (-3σ, $+3\sigma$) about 99.7%. In usual conversation, the first range is simply referred to as "1 sigma", the second "2-sigma", and the third "3-sigma". The larger the sigma number is, the closer the total count to 100% of all possibilities.

Normal approximation of binomial distributions

From our discussion above, when N is large enough, a binomial distribution can be approximated by a normal distribution. For now we will simply state the two relevant formulas for the resulting normal distribution. We will return to the formula for sigma in a later chapter:

Mean, $\mu = Np$ (1)

Sigma, $\sigma = (Nqp)^{1/2}$ (2)

It is customary to refer to p as the "binomial probability for success", meaning the probability of occurrence of the chosen distributive entity in the next trial.

Note: We will eventually abandon probability and refer to p as the "propensity". Since p+q = 1, in calculations we will use (1-p) instead of q.

B. ON THE CONCEPT OF PROBABILITY

The method of doubt

In scientific investigations of controversial subjects it is necessary to apply the "method of doubt" first suggested by the great philosopher scientist René Descartes. This method demands us to explore all possibilities, no matter how remote they appear to be.

On the "ignorance argument" against probability

The traditional argument against the concept of probability is that it is only a substitute for ignorance. Historically, this argument is the main reason why many scientists have been hesitant to accept the Probability theory as a legitimate science.

Typically, the "ignorance argument" starts with the question "Will an event X happen?" There will be cases where, due to lack of information, even the most knowledgeable person in the world cannot answer this question one way or the other.

Logically there are only two possible outcomes: 1."Yes, X will happen" and 2."No, X will not happen". The subjective and classical probability theories say that there is a probability p for X to happen and a probability (1-p) for X not to happen.

This probability answer is presented graphically in the left picture of figure 5. There are two possible outcomes "Yes" and "No", but it is unknown which one will correspond to reality. We tend to assign a probability to each outcome, because both have to be considered to be "in the picture" at the time of prediction. It can be seen that probability is a "reasonable person's reaction" to the unpredictability of future events. All of us tend to make judgments about the unknown future and we tend to express our predictions as probabilities (i.e., replying "possibly 'Yes' or possibly 'No' ".)

The "ignorance argument" points out, however, that what we consider as "unknown" may in fact be deterministic (i.e., known at least to God.) If reality is the right picture of figure 5 then the only correct answer to the question is "No". Thus, at least for this case, the probabilistic prediction of "Yes or No" is no more than a substitute for our ignorance of reality. In particular, the classical probability answer is scientifically wrong because it assigns 50% to each possibility, but reality is 100% for "No".

YES 50%   NO 50%   NO 100%

Probability foundation   Reality
(indeterminate)   (deterministic)

Figure 5: Subjective and classical probability theories state that there is a probability for something to happen and a probability for it not to happen (left picture). The "ignorance argument" appeals to the third possibility, where the information regarding whether an event would happen has already been decided. We are just ignorant of it. Such a case would make the Probability theory a substitute for ignorance instead of a science. As shown here, by choosing "Yes 50% No 50%" the Probability theory is WRONG, because the correct answer happens to be "NO 100%"!!!

It is surprisingly easy to construct all kinds of cases where probability is indeed a substitute for ignorance. Let's look at one of the simplest examples, which happens to be a common way among friends to decide who will pay the bill for lunch. We will flip a coin, cover it on its way down with our hand then ask the question "Heads or tails?" The classical probability answer has to be "50/50". The problem is that at the time this answer is given, the coin is in the deterministic state of either 100% heads or 100% tails, not some probability in between. Thus, the seemingly thoughtful "50/50" statement is indeed a substitute for the ignorant answer "I don't know!"

On the validity of the concept of probability

Now that we have heard the ignorance argument, let's give equal time to the concept of probability.

We will first consider the particular case of a coin flip. We will grant that in extremely rare cases a coin may land on its edge after being flipped. However, we will ignore these cases and concern ourselves only with those cases when the coin settles on one of its sides. For argument's sake, we will assume that the coin is a perfect coin, with no clear tendency toward either heads or tails. While it may be impossible to manufacture such a coin, there is no theoretical problem with this assumption.

We know that after this perfect coin is flipped, it will make complex movements in the air then fall down under the influence of gravity onto whatever surface available to it. When it touches the surface it may bounce and/or roll; but finally it will settle on one of its two sides.

The motions of the coin are so complex that we should not hope to ever analyze them successfully, so we will only focus on the deciding moments before the coin settles on the surface.

The physical shape of the coin has an interesting mechanical consequence. It equalizes all tendencies towards heads, no matter how weak or strong, into "heads"; and it does the same with "tails". In modern language, we say the coin "digitizes" the mechanical impacts that it experiences. With this digitization, the smallest deviation from the balanced position has the same significance as the largest. In fact, a very slight change in conditions (e.g. the vibration of the air) could reverse the outcome from a potential heads to tails, and vice versa.

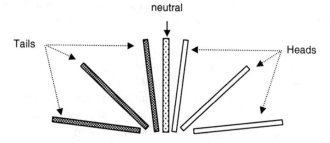

Figure 6: The most interesting property of a falling coin is that it will "digitize" all tendencies to very clear choices of heads and tails; regardless how small or large the tendency is. This digitizing effect makes it impossible to predict the outcome of a coin toss by the deterministic laws of physics.

Since this very fine digitizing action could be arbitrarily small, it removes the last hope that we can ever be successful in predicting the outcome of a coin flip by the method of science, as we know it today. Thus, any deterministic prediction on the outcome of a coin flip will have to appeal to certain "force" still unknown to science.

Can we insist that the outcome of the toss is deterministic, just that we are ignorant of it? We obviously can; but this point is irrelevant because it remains true that we are ignorant of the outcome, regardless of whether it is or is not deterministic.

The same argument applies to the case of a coin that has already settled on one side (but the outcome is hidden from us.) Here we obviously have a deterministic situation, but again this is irrelevant, as it remains true that we are ignorant of the outcome.

The ideal goal of science is to establish certainty, but when certainty cannot be established, what choice is there except to gather all available information, then take a chance with the part that we are ignorant of?

The question therefore is not whether probability is a substitute for ignorance, but whether the Probability theory yields result that fit empirical data. If a Probability theory leads to no verifiable result, it is of little scientific interest (though this does not mean that it is incorrect.) If a Probability theory leads to results that are in agreement with empirical data, it is a viable theory. If a Probability theory leads to results that are in conflict with empirical data, we must question its validity and seriously consider discarding it.

Apparent empirical support for probability

Let's see how frequency and classical probability theories compare with empirical data. In chapter 1 we have presented a set of data that agree with long-term probability prediction. Now that we have introduced sigma, we will apply it to a new test of probability predictions.

In this test we will simulate the averaging of N samples M times, with varying N's and M's. We will choose 3σ as our cut-off point simply because this is the industrial standard at the time of this writing. Frequency probability predicts that the number of trials that deviate outside $\pm 3\sigma$ is approximately $100\%-99.74\% = 0.26\%$ of the total number of trials. Let's see how well this holds up in reality.

Our first task is to confirm that experimental values for sigma agree with theory. For convenience we will define the normalized deviation as:

$$\Delta' \equiv (D-\mu)/\mu = (D-Np)/(Np) \qquad (3)$$

where D is the experimentally recorded occurrences of the event with probability p in each trial of sample size N. Each Δ' is used as a data point to calculate the normalized sigma by known textbook procedure. The calculated normalized sigma is then compared against the theoretical normalized sigma predicted by equation (3).

Combinations of various p's were simulated with sample size N varying from 10 to 120,000 and trial size M between 144 and 240, then tabulated in table 1. By the Central Limit Theorem (which we will cover in a later chapter,) each set of data fits an approximately normal distribution of sample size N (we will see why the relatively small sample size N=10 is also approximately normal in a later chapter.)

As seen on table 1, the actual sigma values agree with probability predictions within +/-10%. We will establish in a later chapter the dependence of sigma on the number of trials. For now we will simply state that for the number of trials given, the agreement is excellent.

TABLE 1: Sigma values – Data versus theory
(test date: August 2001)

Combination	Probability p	Sample size N	Trial size M	Sigma		Error %
				Data	Theory	
1H (1)	0.5	10	144	0.3100	0.3162	-1.97%
1H (1)* - L	0.5	10	240	0.3133	0.3162	-0.92%
1H (1)	0.5	20	144	0.2300	0.2236	2.86%
1H (1)* - L	0.5	20	240	0.2200	0.2236	-1.61%
1H (1)	0.5	200	144	0.0700	0.0707	-1.01%
1H (1)* - L	0.5	200	240	0.0700	0.0707	-1.01%
1H (1)	0.5	4000	200	0.0148	0.0158	-6.19%
1H (1)*	0.5	4000	200	0.0155	0.0158	-2.18%
1H (1)	0.5	40000	200	0.0050	0.0050	-0.67%
1H (1)*	0.5	40000	200	0.0049	0.0050	-1.33%
1H (1)	0.5	120000	201	0.0030	0.0029	5.08%
1H (1)*	0.5	120000	201	0.0027	0.0029	-5.31%
2H (2)	0.25	1500	220	0.0410	0.0447	-8.25%
2T (2)	0.25	1500	220	0.0473	0.0447	5.84%
1H+1T (2)	0.50	1500	220	0.0250	0.0258	-3.30%
1H+1T (2)*	0.50	1500	220	0.0242	0.0258	-6.40%
3H (3)	0.125	2600	240	0.0476	0.0519	-8.20%
3T (3)	0.125	2600	240	0.0556	0.0519	7.09%
2H + 1T (3)	0.375	2600	240	0.0270	0.0253	6.64%
1H + 2T (3)	0.375	2600	240	0.0272	0.0253	7.43%
4H (4)	0.0625	2400	240	0.0863	0.0791	9.20%
4T (4)	0.0625	2400	240	0.0770	0.0791	-2.60%
3H+1T (4)	0.2500	2400	240	0.0333	0.0354	-5.72%
1H+3T (4)	0.2500	2400	240	0.0377	0.0354	6.54%
2H+2T (4)	0.3750	2400	240	0.0257	0.0264	-2.60%
2H+2T (4)*	0.3750	2400	240	0.0270	0.0264	2.46%

2H (2)	0.2500	144,000	201	0.0045	0.0046	-0.68%
3H (3)	0.1250	96,000	201	0.0086	0.0085	0.32%
4H (4)	0.0625	72,000	201	0.0137	0.0144	-5.31%
5H (5)	0.0313	57,000	201	0.0214	0.0233	-8.09%
6H (6)	0.0156	48,000	201	0.0343	0.0362	-5.32%
7H (7)	0.0078	39,000	201	0.0519	0.0571	-8.99%
8H (8)	0.0039	36,000	201	0.0811	0.0842	-3.64%
9H (9)	0.0020	30,000	201	0.1273	0.1305	-2.44%
10H (10)	0.0010	27,000	201	0.1916	0.1947	-1.55%
Note 1: The asterisk (*) denotes repeated experiments for comparison						
Note 2: Experiments with "L" was on Lotus 1-2-3, the rest on Excel						

Next, in table 2 we compare the extreme positive and negative deviations from the ideal average values against 3-sigma to determine whether they are inside the three sigma range (i.e., from −3σ to +3σ).

TABLE 2: Actual deviations vs. 3-sigma's
(test date: August 2001)

			Extreme Deviation from ideal average		3-sigma's compared with data	
	Sample Size	Trial Size			(calculated from raw data)	(Outside 3-sigma)
Combination	N	M	Max %	Min %	3-sigma	Sigma number
1H (1)	10	144	60.00%	-80.00%	93.36%	
1H (1)* - L	*10*	*240*	*80.00%*	*-80.00%*	*94.14%*	
1H (1)	20	144	**70.00%**	-60.00%	69.30%	**3.03**
1H (1)* - L	*20*	*240*	***70.00%***	*-50.00%*	*65.52%*	***3.21***
1H (1)	200	144	13.00%	-21.00%	21.06%	
1H (1)* - L	*200*	*240*	***23.00%***	*-21.00%*	*21.21%*	***3.25***
1H (1)	4000	200	**4.70%**	-4.00%	4.45%	**3.17**
1H (1)*	*4000*	*200*	***4.80%***	***-4.65%***	*4.63%*	***3.11, 3.01***
1H (1)	40000	200	1.26%	-1.18%	1.48%	
1H (1)*	*40000*	*200*	***1.74%***	*-1.12%*	*1.49%*	***3.50***
1H (1)	120000	201	**1.06%**	-0.75%	0.82%	**3.88**
1H (1)*	*120000*	*201*	*0.79%*	*-0.77%*	*0.91%*	

2H (2)	1500	220	10.13%	-11.47%	12.31%	
2T (2)	1500	220	11.73%	-13.60%	14.20%	
1H+1T (2)	1500	220	**8.00%**	-7.20%	7.25%	**3.31**
1H+1T (2)*	*1500*	*220*	*5.87%*	***-8.00%***	7.49%	**3.20**
3H (3)	2600	240	12.62%	-13.23%	14.29%	
3T (3)	2600	240	**18.77%**	-13.85%	16.67%	**3.38**
2H + 1T (3)	2600	240	**10.15%**	-6.87%	8.10%	**3.76**
1H + 2T (3)	2600	240	6.46%	-8.00%	8.16%	
4H (4)	2400	240	**26.70%**	**-27.70%**	25.90%	**3.09, 3.21**
4T (4)	2400	240	20.00%	**-30.70%**	23.10%	**3.99**
3H+1T (4)	2400	240	9.00%	-9.00%	10.00%	
1H+3T (4)	2400	240	**11.70%**	-11.20%	11.30%	**3.11**
2H+2T (4)	2400	240	7.40%	-6.20%	7.70%	
2H+2T (4)*	*2400*	*240*	*6.90%*	***-8.30%***	8.09%	**3.08**
2H (2)	144,000	201	1.35%	**-1.41%**	1.36%	**3.11**
3H (3)	96,000	201	2.37%	-2.41%	2.57%	
4H (4)	72,000	201	3.90%	-4.10%	4.10%	
5H (5)	57,000	201	5.80%	-5.40%	6.43%	
6H (6)	48,000	201	8.40%	-10.10%	10.29%	
7H (7)	39,000	201	**20.10%**	-15.30%	15.58%	**3.87**
8H (8)	36,000	201	**25.20%**	-21.10%	24.33%	**3.11**
9H (9)	30,000	201	33.10%	-31.70%	38.20%	
10H (10)	27,000	201	51.70%	-46.90%	57.49%	

Note 1: The asterisk (*) denotes repeated experiments for comparison

Note 2: Experiments with "L" was on Lotus 1-2-3, the rest on Excel

Of the 70 extreme deviations recorded we find 20 data points (shown in **bold** in table 2) outside the three-sigma range.

The total number of trials for all experiments are:

3(144)+ 4(200) + 11(201) + 4(220) +13(240) =7443

This gives the percentage of trials lying outside the 3 sigma range:

20/7443 = 0.27%

which matches the predicted probability value of 0.26% almost exactly! Thus, our experimental data may be considered as a positive verification for frequency probability, which predicts that in the long run the ratio of events outside three sigma is approximately 0.26% of all events.

Since classical probability claims that the probability for an event outside 3-sigma to happen is 0.26%, the same set of data may also be considered as a positive verification of classical probability.

C. THE LONG-RUN PARADOX

The long-run paradox and the failure of frequency probability

We note, however, that the lowest p value used in our tests so far is 0.0010 (for 10 consecutive heads), which is considered high in the standard of science. If the Probability theory is a consistent description of Nature, it must also work at probability values much lower than 0.0010. In fact, it must work for all probability values. Let's see if this is the case.

For quick reference, we will again show the following test results, which were first reported in chapter 1:

In this test 141.3 billion unbiased tosses were simulated on a Macintosh computer and the consecutive strings of heads were combined in the finally tally. For short strings the frequency probability predictions are well approximated by the actual counts, but significant deviations start to show up with strings longer than 20. Most seriously, strings of length 29 and above are completely absent.

TABLE 3: Computer simulation of 141.3 billion coin tosses
(by Hanspeter Bleuler, Germany, private communication, November 2002)
Counts are combined total for strings that are all-heads

STRING	PROBABILITY PREDICTION	ACTUAL COUNT
1	70,652,392,105	70,650,390,138
2	35,326,196,053	35,326,178,115
3	17,663,098,026	17,663,595,966
4	8,831,549,013	8,832,056,857
5	4,415,774,507	4,416,137,049
6	2,207,887,253	2,208,074,053
7	1,103,943,627	1,104,163,729
8	551,971,813	552,131,003
9	275,985,907	276,011,526
10	137,992,953	137,874,262
11	68,996,477	69,127,258
12	34,498,238	34,545,871
13	17,249,119	17,287,151
14	8,624,560	8,645,262
15	4,312,280	4,311,566
16	2,156,140	2,118,205
17	1,078,070	1,055,064
18	539,035	550,366
19	269,517	265,437

20	134,759	132,277
21	67,379	62,715
22	33,690	34,000
23	16,845	18,727
24	8,422	10,795
25	4,211	2,870
26	2,106	1,309
27	1,053	2,118
28	526	521
29	**263**	**0**
30	**132**	**0**
31	**66**	**0**
32	**33**	**0**
33	**16**	**0**
34	**8**	**0**
35	**4**	**0**
36	**2**	**0**
37	**1**	**0**

Frequency probability predicts that in 141.3 billion tosses, strings of 29 would occur approximately 263 times. We will not go into a detailed calculation, but it suffices to say that the probability to have at least one of these strings to occur is extremely extremely close to certainty. Its absence, therefore, is a serious challenge to the Probability theory.

In hindsight the absence of very long strings is not unexpected. Although the maximum string may vary from computer to computer or/and from software to software, it is mathematically certain that the longest string that each random number generator (e.g., a computer) can produce is finite[1]! Thus, the absence of long strings is a reality in all computing systems; and this fact contradicts probability predictions!

The absence of long strings is sometimes explained by the *ad hoc* concept of a "prohibition threshold", which prevents events with very low probabilities from happening. This is the position of the "Physically Impossible Rationale". We will evaluate the merit of this rationale later in the chapter.

The long-run paradox and the inconsistency of classical probability

Since classical probability equates long-term ratios with event probabilities, it has the same problem as frequency probability at low probability values. In fact, the problem with classical probability is much more serious. To see why this is the case let's return to table 3.

If we apply classical probability to the outcome of the toss immediately following 28 consecutive heads, we will have to conclude that the probabilities for heads and tails are equal at 1/2. However, since an occurrence of heads would make 29 heads in a row, which is a prohibited event, in reality the probabilities for heads and tails are 0 and 1 respectively:

Probability	Classical probability	Reality
P(H)	1/2	0
P(T)	1/2	1

If we apply classical probability to the next two tosses after 27 consecutive heads have just occurred, we will have to predict that HH, HT, TH, TT are equally probable at P=1/4. However, since HH would make 29 heads in a row, it is clear that it cannot occur, leaving only HT, TH, and TT as realizable outcomes. Since the probability for HH is 0, the total probabilities for the other three possibilities must add to 1 (versus the incorrect value of 3/4 predicted by classical probability).

How should we deal with the other three combinations? It sounds reasonable to split 1 equally and assign the probability value of 1/3 to each of them. The problem is, we have no way to know if this arrangement is logical. But with simple accounting we do know one thing for sure: The correct probability values, whatever they are, will not match the classical probability values!!!

Probability	Classical prediction	Reality
P(HH)	1/4	0
P(HT)	1/4	1/3 (?)
P(TH)	1/4	1/3 (?)
P(TT)	1/4	1/3 (?)

By continuing with this logic for the next 3 tosses after 26 consecutive heads, the next 4 tosses after 25 consecutive heads, etc., up to the next 28 tosses after a single heads occurrence, we are led to the conclusion that event probability must vary from toss to toss, which contradicts classical probability.

Thus, by *reductio ad absurdum*, classical probability is either incorrect or incompatible with computer operations.

The Infinite Randomness Assumption
and its failure in man-made computing systems

We know from chapter 1 that the Probability theory assumes that all distributive processes are ruled by total randomness; meaning that there exists no element of determinism, no matter how minute. For reference purposes, we will call this "The Infinite Randomness Assumption".

Since a computer was used as a random number generator to obtain the data in table 3, and it is known that the "randomness capacity" of all man-made random number generators is finite[1], one way to defend classical probability is to blame the computer for its failure to meet the Infinite Randomness Assumption.

We will evaluate the general validity of the Infinite Randomness Assumption in the next chapter. However, the fact that this assumption is not met by the computer -which has been extremely successful in simulating real life distributive processes- is a serious step backward for science. If we try to hang on to the Probability theory, we will have to make computer operation an exception to the rule. This is hardly a satisfactory solution because it suggests the obvious but very troubling question: "What if more exceptions show up in the future?" This question forces us to take a critical look at the Probability theory.

D. THE "PHYSICALLY IMPOSSIBLE RATIONALE" AND THE LAW OF LARGE NUMBER

The "Physically Impossible Rationale" and the reason why it contradicts the Probability theory

There is one argument sometimes used by probability proponents to explain why events with very low probabilities (e.g., very long strings of heads or tails in coin tossing) have never been observed. The argument goes like this: "Events with very low probabilities are mathematically possible but physically impossible." We will call this the "Physically Impossible Rationale". According to the Physically Impossible Rationale, there exists a "prohibition threshold". Events with probabilities below the prohibition threshold cannot occur.

Although the "Physically Impossible Rationale" is quite popular with engineers who deal with distributive processes on regular basis, the writer is not aware of any official argument for it in scientific literature. This limits its status to an empirical law. But at any case, there are many examples where the physical impossible rationale appears necessary. The following example is taken from thermodynamics.

Let's assume we have two compartments that contain gases A and B respectively and are separated by an impenetrable barrier. For simplicity, A and B will be assumed to be non-reacting. It is well known that if the impenetrable barrier is removed, A and B will mix together. This mixing is explained as a probabilistic result. The mixed state is realized because it is the state of highest probability[2].

The trouble is, also by probabilistic calculations, we find that there is a non-zero, albeit minute, probability that the two gases do not mix at all. If the standard interpretation of probability is applied, we would be forced to accept that it is possible, though extremely unlikely, to have a case of no mixing!

SYMMETRY AND THE END OF PROBABILITY

In fact, there is a finite probability that, after complete mixing has taken place, the system will return to the unmixed state if given enough time. This was first shown by Henri Poincaré in 1890 (then again in 1893) and is now known as Poincaré's Recurrence Theorem[2].

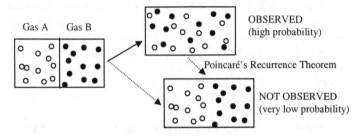

Figure 7: We start with a two-compartment chamber, with non-reacting gases A and B separated by an impenetrable barrier (left). When the barrier is removed, the two gases always mix together (top right). The unmixed state (bottom right) has never been observed. The probability theory argues that the mixed state is realized because its probability is high, the unmixed state not realized because its probability is extremely low. The problem is, even an extremely low probability could happen. Many engineers have chosen to resolve this issue by taking the position that events with very low probabilities cannot take place. This is the "Physically Impossible Rationale".

Then why gases always evolve to the mixed state and stay there? There are two possible answers. The first answer is popular among scientists. It dodges the question "Why there has never been a case of the unmixed state?" by accepting the possibility that the unmixed state may occur any time in the future, but the probability for such an event is extremely low. Needless to say, this answer leaves science in a very shaky position (because the event that does occur may also have very low probability.)

The second answer is popular among engineers, who are a practical lot. It rules out the case of non-mixing by arguing that it is physically impossible for events with very low probabilities to take place. This is the essence of the "Physically Impossible Rationale".

Let's apply the "Physically Impossible Rationale" to the case of coin tossing. If the single event probability to get heads is p, then the probability $P(N/N)$ to get a string of N consecutive heads is:

$$P(N/N) = p^N \qquad (4)$$

Solving for N:

$$N = \ln[P(N/N)] / \ln[p] \qquad (5)$$

where $\ln[x]$ is the logarithm of x. Let P_{min} be the prohibition threshold. It is clear that $P(N/N)$ cannot be smaller than P_{min}. The longest string possible N_{max}, then, can be approximated by replacing N by N_{max} and $P_{(N/N)}$ by P_{min} in equation (5):

$$N_{max} \approx \ln[P_{min}] / \ln[p] \qquad (6)$$

Since P_{min} is non-zero, N_{max} has to be finite, and the long-run paradox is solved. This is why the Physically Impossible Rationale is attractive. The bad news is, the Physically Impossible Rationale contradicts the basic concept of probability, as we will see shortly.

Figure 8 is the situation for 4 unbiased coins. We will let "H" stand for heads, and "T" for tails. According to the probability theory, the combination 2H2T are 6 times more abundant than 4H and 4T in the long-term distribution not because nature is biased toward 2H2T, but because 2H2T has 6 equivalent combinations in stock versus one for each of 4H and 4T. As far as the selection process is concerned, one combination will be chosen each time from the stock.

"EQUAL OPPORTUNITY" COMBINATIONS IN STOCK

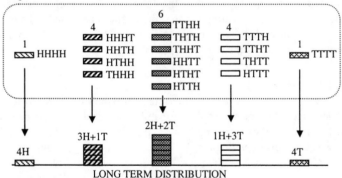

Figure 8: The probability picture of a distribution of 4 unbiased coins. According to this picture, all combinations are equivalent. The reason that four heads and four tails occur less frequently is believed to be that each only has one combination (HHHH and TTTT), while others have more. This description is in conflict with the reality of prohibited propensity, because if one combination (e.g., HHHH) is prohibited, then the others are also prohibited, and none of them can ever occur, which is an absurdity.

The probability for both HHHH and TTTT is $0.5^4 = .0625$. What if this probability is below the prohibition threshold and therefore is prohibited from occurring? We would attempt to answer that only the other 14 combinations will occur, and the equal opportunity principle still applies separately to each of them. According to the Probability theory, however, there is no difference between HHHH or TTTT and any of the other 14 combinations. If HHHH and TTTT are prohibited, it would follow that the other combinations are also prohibited and nothing can happen; which is an absurdity.

Since HHHH and TTTT are prohibited but the other combinations are not, we conclude that the "Physically Impossible Rationale" indeed contradicts the Probability theory.

The following alternative proofs were included for readers who love logic puzzles:

ALTERNATIVE PROOF 1:

Let P_{min} be the prohibition threshold and p the probability of the prohibited event. Clearly $p < P_{min}$ and that's why the event cannot take place. Note, however, that the probability for the event to occur at least once in N attempts increases with N according to the formula

$$P(\geq 1/N) = 1-(1-p)^N \qquad (6)$$

By choosing an arbitrarily large N, we can make $P(\geq 1/N)$ not only larger than P_{min}, but also as close to 1 as we please. This means the prohibited event should occur in the long run. However, this would contradict the starting assumption that the event is prohibited. Thus, by *reductio ad absurdum*, we conclude that the "Physically Impossible Rationale" contradicts the theory of probability.

ALTERNATIVE PROOF 2:

We will conduct a thought experiment in which N unbiased tosses are attempted. We know that the probability for each possible ordered sequence is 0.5^N. By increasing N we can make this probability smaller P_{min}. Since an event with probability smaller than P_{min} cannot occur, none of the sequences can occur; which is an absurdity. Again, by *reductio ad absurdum*, we conclude that the "Physically Impossible Rationale" contradicts the theory of probability.

The Law of Large Number
and the reason why it is in conflict with the Probability theory

Some probability proponents may argue that the peculiar behavior of computers as cited can still be accounted for by the theory of probability, by applying its most well known result, namely the Law of Large Number.

For readers who are not familiar with the concept, the Law of Large Number states that, when we take N samples from a distribution, by increasing N we can make the average value of these N samples as close as we desire to the average value of the distribution.

The Law of Large Number has long been regarded as a major theoretical triumph of the probability theory. As it turns out, this is an unjustified claim based on an invalid proof. It can be counter-proved that the Law of Large Number has nothing to do with the probability theory, and in fact contradicts it. The proof for this is in the appendix following this chapter. Readers who love mathematical proofs may want to read it first before continuing with this chapter. Readers who are not comfortable with mathematical proofs can still grasp the essence of the logic by reading the rest of this section.

Although in the Law of Large Number N is understood to be a very large number, we can use a case of small N to expose the logical problem hidden in the textbook proof of the Law of Large Number. Take the case N=3 in the process of unbiased coin tossing. To facilitate the argument, we will arbitrarily set heads=1 tails=0.

Since the probabilities to get heads and to get tails are equal at 1/2, and each event involves three tosses, the classical probability for each of the 8 possible events is equal to:

$1/2 \times 1/2 \times 1/2 = 1/8$

This means the 8 possible events are equivalent and distinct. Each is as likely as the others to be the next event (see last column of table 4).

TABLE 4: Averages and probabilities in 3 coin toss
Groups in B: [1], [2,3,4], [5,6,7], [8]

Possibility	Sequence	Total	Average	Probability
1	HHH	3	1	1/8
2	THH	2	2/3	1/8
3	HTH	2	2/3	1/8
4	HHT	2	2/3	1/8
5	TTH	1	1/3	1/8
6	THT	1	1/3	1/8
7	HTT	1	1/3	1/8
8	TTT	0	0	1/8
Overall			0.5	1.0

However, according to the Law of Large Number, possibilities [2,3,4] are indistinguishable (considered as a single possibility) because they all have the same average of 2/3; possibilities [5,6,7] are also indistinguishable because they all have the same value of 1/3. Possibilities [1] and [8] stand alone because they are the only one with average 1 and 0, respectively. *(See Appendix at end of chapter).*

It is clear that the Law of Large Number and the probability theory operate with different criteria; which we will call "the probability criterion" and "the average criterion", respectively. It can be shown that the probability criterion and the average criterion conflict each other *(see appendix.)* Since even a single conflict is not acceptable in mathematical proofs, the existing proof of the Law of Large Number is invalid. The shocking conclusion is: *The Law of Large Number CANNOT be a result of the probability hypothesis.*

Further –as shown in the appendix- conflicts can be avoided in the proof for the Law of Large Number by discarding the probability interpretation of distributions and return distributions to their original meaning, i.e., distributions and not probabilities. This implies that *The Law of Large Number is a property of distributions, and has nothing to do with probabilities.*

Since the Law of Large Number is based on the average value, for identification purposes we will call this "the average criterion of the Law of Large Number". The puzzling question is: What is the justification for the average criterion? We will address this question in the next chapter.

Since both the Law of Large Number and the Physically Impossible Rationale fit reality, their disagreement with the Probability theory is a clear indication that the Probability theory has very serious problems at the very fundamental level.

E. THE GIBBS PARADOX

The Gibbs paradox as another failure of probability

We have mentioned the scientific field of "thermodynamics" in the gas mixing problem. It is thermodynamics that gave us the buzzword "entropy", which is a measure of how random a system is. The "law of entropy", also known as "the second law of thermodynamics", has become a somewhat popular notion, and we do hear it sometimes in social conversations. In simple terms, the law of entropy says that the total entropy of a system isolated from the external world never decreases (and often increases.) The calculations for entropy of gas mixing makes use of a quantity called the partition function (the reader does not need to know what this function is.) The partition function for ideal gas was derived by Ludwig Boltzmann by the method of probability.

Again we look at a chamber whose two compartments are separated by an impenetrable barrier; but now we are interested in the entropy increase due to the gas mixing process when the barrier is removed. When one compartment is filled with gas A, the other with gas B, the mixing should cause an increase in entropy because the mixed state is not as well ordered as the unmixed state (top picture of figure 9). Since ΔS is the conventional notation for entropy increase, we have $\Delta S > 0$; which agrees with the prediction given by the Probability theory.

The paradoxes of probability

Now consider the trivial case with gas A occupying both compartments under identical conditions. When the barrier is removed it is obvious that there is no increase in randomness, therefore $\Delta S=0$ (bottom picture of figure 9). The problem is, the probability prediction gives $\Delta S>0$. This is known as the Gibbs paradox.

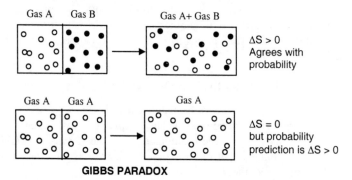

GIBBS PARADOX

Figure 9: The mixing of two different gases (top) gives an entropy increase, as expected. The mixing of the same gas should give no entropy increase (bottom). However, the probability theory still predicts an entropy increase. This is known as the Gibbs paradox. The Gibbs paradox is easily solved by recognizing that the gas atoms are equivalent. The problem is, this recognition process is not available in the framework of probability. This means the Probability theory is either wrong or incomplete.

The Gibbs paradox is solved by recognizing that, as far as the overall picture is concerned, the situations before and after the removal of the barrier are identical when both compartments contain the same gas. This solution is of course correct, but it is obviously in conflict with the concept of probability. To demonstrate this conflict let's consider 3 individual gas atoms X, Y, Z and three positions 1, 2, and 3 available to them. Among XYZ, there are 6 possible arrangements with equal probabilities:

Arrangement	X	Y	Z
I	1	2	3
II	1	3	2
III	2	1	3
IV	2	3	1
V	3	1	2
VI	3	2	1

From a probability standpoint, although these 6 arrangements are equivalent; they have to be considered distinct because probabilistic events are supposedly independent of one another. However, in the solution of the Gibbs paradox, the 6 arrangements in the example are considered as *a single arrangement* because the "system" cannot distinguish them. The alert reader will notice that this "global indistinguishability" is a manifestation of the same "average criterion" employed by the Law of Large Number.

Since the average criterion is in conflict with the probability criterion, the Gibbs paradox is another indication that the probability criterion, and hence the Probability theory, is either wrong or incomplete.

F. THE PARADOXES OF PROBABILITY

We have presented two major probability paradoxes:
1. The long-run paradox can be explained by either the "Physically Impossible Rationale" or the Law of Large Number; but both of these explanations are incompatible with the Probability theory.
2. The Gibbs paradox can be explained by recognizing that there is no overall change when two volumes of the same gas are mixed together. The problem is, this solution is incompatible with probability, which must analyze the problem from the standpoint of the individual atoms.

These paradoxes imply that the Probability theory is incompatible with real life distributive processes. The inevitable conclusion is that either the theory is wrong or it has left out something very important.

It is also clear that a scientifically acceptable theory for distributive processes must be consistent with the Law of Large Number and the Physically Impossible Rationale. We will start the search for such a theory in the next chapter. If our search is not successful, we have no choice but to stay with the Probability theory and accept its drawbacks; but if our search is successful we must choose the new theory and abandon the Probability theory.

First written January 2001
Revised August 2001, November 2001, December 2002

NOTES

1. Finite randomness capacity of computing systems

There are two theorems of computational science. The first theorem states that it is impossible to construct a system that possesses infinite randomness. The second states that a system can never produce more randomness that it possesses. These two theorem together implies that all computers, including the most advanced computer that man can ever create in the distant future, have a finite "randomness capacity" and therefore can never produce an endless series

2. Poincaré and the probability prediction of gas mixing

An interesting historical note: It was Poincaré who put thermodynamics in trouble with his Recurrence Theorem (1890). Interestingly, it was again Poincaré who came to its rescue with the Poincaré Theorem (1898) which, for example, explains why gases mix together. The logic is that there are more mixed states available than unmixed states, therefore the probability for the mixed state is much higher than that of the unmixed state.

A quick inspection reveals that this "probability prediction" by Poincaré Theorem is an application of the Law of Large Number. Since the Law of Large Number has been shown in this chapter to be in conflict with the theory of probability, the so-called "probability prediction" is a case of mistaken identity.

3. Subjective probability and propensity

When the task is to predict the outcome of a single event, the only meaningful answers are "yes", "no", and "undecided". For this reason, the following statements would be completely useless:

1. "The probability for 'yes' is 0.5", which is a classical probability statement.
2. "If many events are attempted, the ratio for 'yes' will be approximately 0.5" (which is a common statement for both classical and frequency probability).

This shows clearly that classical probability and frequency probability are not the appropriate tools for single event prediction. We are left with subjective probability and possibly Karl Popper's propensity interpretation. Although these views have great philosophical implications, they are of little interest to existing science because they generally do not yield testable results.

4. The view of classical probability on single events and series of events

Classical probability maintains that the probabilities for all single events in a process are identical (and equal to the long-term average of the distribution.) This is equivalent to saying the best and the worst students in a class are expected to have the (same) average score in the next examination; which we know to be incorrect.

It is therefore important to realize that an individual event and the long-term distribution of a process may not have anything to do with each other. This does not mean that we cannot analyze the individual events. It only means that such an analysis is outside the scope of any theory that chooses the whole process as its subject of study. Classical probability is one such theory.

APPENDIX I

On the reason why the Law of Large Number has nothing to do with the Probability theory and in fact contradicts it

Modern proof of the Law of Large Number
(based on Chebyshev's inequality)

The textbook probability proof for the Law of Large Number typically starts with the Chebyshev's inequality, which has many known versions. We will use the following version:

"The probability P that an actual value X of a sample taken from a given distribution is outside the range $(\mu - \varepsilon, \mu + \varepsilon)$, where μ is the average value of the distribution and ε an arbitrarily chosen small number, is less than σ^2/ε^2, where σ^2 is the variance of the distribution being investigated."

The proof for this version is as follows:

By definition: $\sigma^2 \equiv \Sigma (X_i - \mu)^2 P_i$ (1a)

Therefore: $\sigma^2 > \Sigma^*(X_i - \mu)^2 P_i$ (2a)

where only those X_i's that meet the condition $| X_i - \mu | > \varepsilon$ are included in Σ^*.

Obviously: $\sigma^2 > \Sigma^* \varepsilon^2 P_i = \varepsilon^2 \Sigma^* P_i$ (3a)

But: $\Sigma^* P_i = \Sigma^* P(X=X_i) = P(| X_i - \mu | > \varepsilon)$ (4a)

Hence: $\sigma^2 > \varepsilon^2 P(| X_i - \mu | > \varepsilon)$ (5a)

Finally: $P(|X - \mu| > \varepsilon) < \sigma^2/\varepsilon^2$ (6a)

Following are standard steps of the proof for the Law of Large Number, as found in textbooks.

1. Given a distribution A which possesses a finite average value μ_A and a finite variance σ_A^2. Consider distribution B formed by the average of N samples at a time from distribution A (i.e., each set of N randomly chosen samples of distribution A is transformed to a single data point in distribution B.)

2. Let X_{Ai} be the ith sample in the group of N samples taken from distribution A, and X_B the value of the data point representing these N samples in distribution B.

3. Clearly:
$X_B = \Sigma X_{Ai} / N$ (7a)
where the summation Σ is for all i's from 1 to N.

4. It can be shown:
$\mu_B = \mu_A$ (8a)
i.e., the mean value of distribution B is the same as that of distribution A.

5. It can be shown:
$\sigma_B^2 = \sigma_A^2/N$ (9a)

6. Apply Chebyshev's inequality (6a) to distribution B we get:

$$P(|X_B-\mu_B|>\varepsilon) < \sigma_B^2/\varepsilon^2 \tag{10a}$$

7. Substituting (8a) and (9a) into (10a):
$$P(|X_B-\mu_A|>\varepsilon) < \sigma_A^2/(N\varepsilon^2) \tag{11a}$$

8. Inequality (11a) can be rewritten as:
$$P(|X_B-\mu_A|\leq\varepsilon) >1- \sigma_A^2/(N\varepsilon^2) \tag{12a}$$

Since σ_A and ε are fixed, $P(|X_B-\mu_A|\leq\varepsilon)$ converges to 1 when N approaches infinity. In words: "The probability that X_B is arbitrarily close to the mean value approaches 1 when N approaches infinity".

This completes the proof for the Law of Large Number, or more precisely the Weak Law of Large Number. As the reader may guess, there is also a Strong Law of Large Number. It turns out that both Laws of Large Number apply to the average value of samples taken from an original distribution. We will argue that because of this property the Weak Law of Large Number cannot be a result of the Probability theory. The same conclusion then automatically applies to the Strong Law of Large Number.

It is therefore sufficient to discuss only the Weak Law of Large Number; which we will call "the Law of Large Number" for brevity. At a quick glance, the proof for the Law of Large Number seems flawless. We will argue, however, that it relies on an implicit condition which cannot be derived from the Probability theory.

That the Law of Large Number
may not be a logical result of the Probability theory

At a quick glance, the proof for the Law of Large Number seems flawless. We will argue, however, that it relies on an implicit condition which cannot be derived from the Probability theory.

Since there is no particular restriction on distribution A, we can apply (12a) to the particular case where A is the distribution of heads and tails in a coin tossing process. The reason we choose coin tossing is because it is familiar to most readers. The logic is easily extended to other distributions. For mathematical convenience we will give heads the value 1 and tails the value 0.

Although the sample size N implied in the Law of Large Number is large, we can use a small N to demonstrate the implication of this classification method without any loss of generality. We will choose N=3 and consider the particular process of tossing 3 unbiased coins. The 8 possible sequences are:

HHH, HHT, HTH, THH, HTT, THT, TTH, TTT

According to the Probability theory, these 8 possibilities are equivalent (because the probability is $1/2^3=1/8$ for each) but distinguishable (because each is a unique sequence). We will call this the "probability criterion".

By definition B is a distribution of the average value of N data points taken from A. This means sequences in B that have the same average value are indistinguishable (identical to one another), and sequences (in B) that do not have the same average are distinguishable (different from one another).

There are only four distinguishable groups in B:
3H, 2H1T, 1H2T, 3T

More specifically:

HHT, HTH, THH are all considered as 2H1T, therefore the probability for one of them to occur is 3×1/8=3/8.

TTH, THT, HTT are all considered as 1H2T, therefore the probability for one of them to occur is 3×1/8=3/8.

HHH has no other equivalent, therefore the probability for it to occur is 1×1/8=1/8.

TTT has no other equivalent, therefore the probability for it to occur is 1×1/8=1/8.

This means, in the classification method applied to distribution B, the three sequences HHT, HTH, THH are allowed to add their individual probabilities together to improve the chance for at least one of them to occur. Sequences TTH, THT, HTT have the same privilege. Sequences HHH and TTT, for having no other sequence with the same average, are left out. We will call this criterion "the average criterion", because it classifies sequences according to their averages and their averages alone.

Surprisingly, the average criterion is taken for granted in the construction of distribution B without any justification at all.

The situation is summarized in table 1A.

TABLE 1A: Eight possible sequences with 3 unbiased tosses
Distinguishable groups in B: [1], [2,3,4], [5,6,7], [8]

Possibility	Sequence	Total	Average	Classical Probability
1	HHH	3	1	1/8
2	THH	2	2/3	1/8
3	HTH	2	2/3	1/8
4	HHT	2	2/3	1/8
5	TTH	1	1/3	1/8
6	THT	1	1/3	1/8
7	HTT	1	1/3	1/8
8	TTT	0	0	1/8
Overall			0.5	1.0

The reader should convince him or herself that the average criterion cannot be derived from the classical probability criterion. This means two independent criteria are used in the existing proof for the Law of Large Number: The classical probability criterion (to establish the Chebyshev's inequality) and the average criterion (to define distribution B).

If we denote the classical probability criterion by CP, the average criterion by X, and the Law of Large Number by LLN; then the existing proof for the Law of Large Number is of the form:

CP & X => LLN   (13a)

Since this is neither CP => LLN nor CP⇔LLN, the Law of Large Number may not be a logical result of classical probability.

Symmetry and the end of probability

That the existing proof for the Law of Large Number is internally inconsistent and therefore logically invalid

We will now argue that the probability criterion and the average criterion are in conflict with each other.

Although the effect is significant only at large sample size N, we can conduct the same analysis with small sample size to demonstrate the problem. Let's take a particular situation where 2 unbiased tosses have turned up heads and we are about to attempt the third toss. Details are shown in table 2A.

TABLE 2A: Outcomes of third toss and resulting averages

Case	First two	Third	Total	Average
a	2H	H	3H	1
b	2H	T	2H1T	2/3

According to the probability criterion, the outcome of the third toss is independent of the outcomes of the first two tosses. Since the process is unbiased we must conclude that, as far as the third toss is concerned, heads and tails are equivalent.

We note, however, that the average for all three tosses will be 1 if the third toss is heads, and 2/3 if the third toss is tails. Since the averages for the two cases are different, according to the average criterion we must conclude that, as far as the third toss is concerned, heads and tails cannot be equivalent!!!

Thus, we have established a case where the probability criterion and the average criterion conflict each other. The reader can verify that this conflict exists in the general case: The probability criterion claims that future outcomes are indifferent to previous outcomes, but the average criterion claims otherwise!

Since the existing proof for the Law of Large Number uses both criteria, it follows that the proof itself must be internally inconsistent; which is another way to say that it is logically invalid.

That the Law of Large Number is a distribution property and has nothing to do with the Probability theory

Since the probability criterion and the average criterion are in conflict with each other, to be logically consistent we can only keep one of the two. If we keep the probability criterion all we have is the Chebyshev's inequality. This will not be a sufficient condition to prove the Law of Large Number because there is no information regarding distribution B.

If we keep the average criterion, we have all we need to know about distribution B. The Law of Large Number can be proven if we can find a non-probabilistic relationship equivalent to the Chebyshev's inequality. This is easy because the Chebyshev's inequality as used in the existing proof for the Law of Large Number is only the probabilistic interpretation of a more natural version of the same inequality, which applies to distributions. The natural version of the Chebyshev's inequality can be stated as follows:

"The collective distribution density D of samples -taken from a given (normalized) distribution- lying outside the range $(\mu - \varepsilon, \mu + \varepsilon)$, where μ is the average value of the distribution and ε an arbitrarily chosen small number, is less than σ^2/ε^2, where σ^2 is the variance of the distribution being investigated."

The proof for this version is as follows:

By definition: $\sigma^2 \equiv \Sigma (X_i - \mu)^2 D_i$ (1b)

Therefore: $\sigma^2 > \Sigma^*(X_i - \mu)^2 D_i$ (2b)

where only those X_i's that meet the condition $| X_i - \mu | > \varepsilon$ are included in Σ^*.

Obviously: $\sigma^2 > \Sigma^* \varepsilon^2 D_i = \varepsilon^2 \Sigma^* D_i$ (3b)

But: $\Sigma^* D_i = \Sigma^* D(X=X_i) = D(| X_i - \mu | > \varepsilon)$ (4b)

Hence: $\sigma^2 > \varepsilon^2 D(| X_i - \mu | > \varepsilon)$ (5b)

Finally: $D(|X-\mu| > \varepsilon) < \sigma^2/\varepsilon^2$ (6b)

We will now use this version of the Chebyshev's inequality to prove the Law of Large Number:

1. Given a distribution A which possesses a finite average value μ_A and a finite variance σ_A^2. Consider distribution B formed by the average of N samples at a time from distribution A (i.e., each set of N randomly chosen samples of distribution A is transformed to a single data point in distribution B.)

2. Let X_{Ai} be the ith sample in the group of N samples taken from distribution A, and X_B the value of the data point representing these N samples in distribution B.

3. Clearly:
$X_B = \Sigma X_{Ai} / N$ (7b)
where the summation Σ is for all i's from 1 to N.

4. It can be shown:
$\mu_B = \mu_A$ (8b)
i.e., the mean value of distribution B is the same as that of distribution A.

5. It can be shown:
$\sigma_B^2 = \sigma_A^2/N$ (9b)

6. Apply Chebyshev's inequality (6b) to distribution B we get:
$D(|X_B - \mu_B| > \varepsilon) < \sigma_B^2/\varepsilon^2$ (10b)

7. Substituting (8b) and (9b) into (10b):
$D(|X_B - \mu_A| > \varepsilon) < \sigma_A^2/(N\varepsilon^2)$ (11b)

8. Inequality (11b) can be rewritten as:
$D(|X_B - \mu_A| \leq \varepsilon) > 1 - \sigma_A^2/(N\varepsilon^2)$ (12b)

Since σ_A and ε are fixed, $D(|X_B - \mu_A| \leq \varepsilon)$ converges to 1 when N approaches infinity. Since the total area under the distribution curve adds to 1, it follows that every X_B separately satisfies $|X_B - \mu_A| \leq \varepsilon$. In words: "$X_B$ is arbitrarily close to the mean value when N approaches infinity".

This completes our new proof for the Law of Large Number. Since the logic for the new proof is parallel to that of the existing proof, its construction is at least equally valid. The big difference is that the new proof is not based on probability. The contradiction between the probability criterion and the average criterion does not exist; and the new proof is perfectly valid.

However, the alert reader will notice that the new proof has not addressed an intriguing question: "What is the logical justification for the average criterion?" Actually this question applies equally to both proofs, i.e., the existing proof based on probability and the new proof based on distribution. It was not raised in the existing proof because no one recognized the (implicit) existence of the average criterion.

We will provide an answer to this important question in the next chapter.

APPENDIX II

The Chebyshev's inequality and why statistical distributions cannot be interpreted as "probability density functions"

A statistical distribution is the long-term pattern formed by a large enough number of individual events in a distributive process. The larger the number of events, the more well defined the distribution. The most familiar statistical distribution is the normal curve representing many coin tossing trials of N coins each, with N being a very large number.

As evident in the probabilistic proof for the Chebyshev's inequality (appendix I), the Probability theory takes for granted that statistical distributions can be interpreted as "probability density functions". The validity of this practice has never been questioned, but we will now re-examine it in detail.

We will start with an arbitrary looking statistical distribution, which we present graphically in figure 1A with "value" on the horizontal axis and "frequency" on the vertical axis. Following the Probability theory, we will divide the area the curve F by itself (without changing the horizontal values). This operation gives the normalized statistical distribution curve F', which is identical to "probability density function" P.

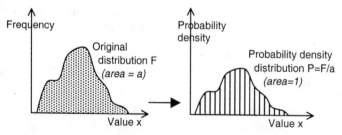

Figure 1A: The transformation of statistical distribution (left) to probability density function (right). Note that the total area under the probability density function P is exactly 1.

Let's attempt to apply the probabilistic Law of Large Number to an arbitrary segment Δx_i of the normalized statistical distribution. If the number of trials is finite but very large, the probability P_0 that F' is confined within a reasonable range (i.e., $LL_i \leq F'_i \leq UL_i$) can be assumed to be very close to 1.

SYMMETRY AND THE END OF PROBABILITY

$P_0 \approx 1$ (1c)

This probabilistic reasoning seems to have successful accounted for the routine reproducibility of statistical distributions in practice; but we will now show that this view is a logical mistake.

The area under P is exactly 1. This means, if we divide the horizontal range to n equal segments of length Δx so that $n\Delta x = x_0 - x_n$, where x_0 and x_n are the minimum and maximum values respectively, and define $P_i = P(\Delta x_i)$ as the probability that x is between x_{i-1} and x_i, then the following must hold true:

$\Sigma P_i = P_1 + P_2 + P_3 + \ldots + P_n = 1$ (2c)

It follows from (1c) and Kolmogorov's axiom 2 (chapter 1, page 11) that the individual probabilities presented by vertical sub-areas in the left picture of figure 2A are independent of one another. Since the probability density curve is but an image of the statistical distribution, each segment of the normalized statistical distribution must also be independent of other segments.

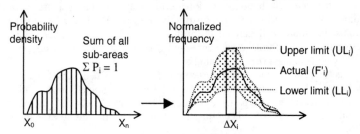

Figure 2A: In the formalism of the Probability theory, the formation of statistical distributions is also probabilistic. Since statistical distributions are routinely reproduced, every point of an actual distribution must be within a finite range. It can be shown that the probabilistic explanation for this reproducibility of statistical distributions is based on a fallacy.

Since ΔX_i can be divided into m sub-segments, which also have to be considered independent of one another, P_0 must also obey:

$P_0 \leq (P_{max})^m$ (3c)

where P_{max} is the highest probability recorded for the set of m sub-segments derived from ΔX_i. Since the number of trials is finite $P_{max} < 1$. With m being arbitrarily large (3c) implies:

$P_0 \approx 0$ (4c)

Since (4c) is in conflict with (1c), by contradiction it is clear that individual probabilities in a probability density function cannot be independent of one another. This invalidates the interpretation of normalized statistical distributions as probability density functions. Since this interpretation is a fundamental part of the Probability theory, we are forced once more to conclude that the Probability theory has been based on a fallacy.

Chapter 3

Symmetry
and the End of Probability

We will adopt the Popperian concept of propensity, with proper modifications, as the description for distributive processes.

We will show that randomness is opposed by the force of non-randomness. We will argue that the force of non-randomness owes its root to symmetry. We will show that the long-run paradox and the Gibbs paradox of the Probability theory can be easily solved by applying symmetry.

Since symmetry contradicts the theory of Probability. Our choice of symmetry marks the End of Probability.

A. MORE REVIEWS OF STATISTICAL CONCEPTS

Replacement of "probability" by "propensity"

In anticipation of a permanent departure from the Probability theory, we will attempt to replace the terms "probability" and "probability value" by a more meaningful term.

The word "propensity" immediately comes to mind because this Popperian concept reminds us, for example, that an unbiased coin may give a long-term ratio of say 0.9 for heads if it is tossed by a well programmed robot. Since we have been conditioned to believe that the probability for heads is 0.5 for such a coin, it would be more meaningful to use "propensity" to refer to the long-term ratio of 0.9.

Unfortunately Karl Popper has taken the position that all events with the same experimental condition will have the same "propensity"[1]. This makes the concept of "propensity" no more than a refined version of "classical probability". For this reason, this concept will also fail to account for the absence of events with very low "propensities".

To avoid confusion, the writer reluctantly chose the term "long-term bias" (often abbreviated as "bias") to replace "probability" in the first edition of this book. Now that more than one year has passed, with more careful thinking, he has a new thought that he would like to share with the reader. His thought is as follows:

"The incompleteness of an idea should not stop us from respecting its insightfulness; and Popper's idea of propensity was indeed insightful."

With this thought in mind, the writer will make the unusual and rather complicated proposal that we will use the term "propensity" to replace "probability", with the condition that the term "propensity" is understood in its post-Popperian meaning, as a long-term tendency in a long enough series of events, not as the tendency of a single event.

A more in-depth discussion of the concept of "propensity" can be found in the appendix at the end of this chapter.

Sample size and trial size

Up to now, when we flip a coin N times we tend to say that the number of trials is N. This would create a problem if we flip N coins M times because we would have to say "M trials, each trial with N trials"; which would be very confusing. A better way to describe this situation is to consider the repeating N "trials" as a unit call "sample size" and say "M trials of N samples" or "M trials of sample size N".

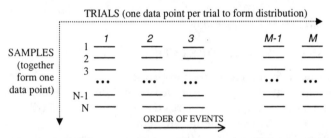

Figure 1: When we repeat a test N times and combine them into 1 result (one data point) we say our sample size is N. To obtain many data points we may want to repeat the same set of N samples many times to get M trials.

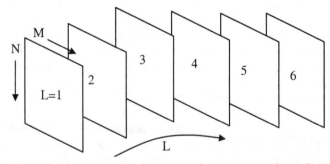

Figure 2: There is no rule that says we have to stop at the trials M. We can repeat many trials to get the "double-trials" L's. This process could go on indefinitely to K's, J's, I's, etc. The choice of alphabetical order is curious (backward in time order) simply because someone has chosen N as the sample size long time ago and we are stuck with this choice.

This differentiation will be very helpful later when we apply the Central Limit Theorem, which involves M trials of sample size N. Each unit of N samples only counts as one data point, and the resulting (approximately normal) distribution will have a total of M data points.

We will call N the sample size and M the trial size.

Normal approximation of binomial distribution (review)

This section is intended to be a quick reference on the basics of normal approximation of binomial distribution.

Recall from chapter 2 that for trials of sample size N, if N is large enough the situation can be approximated by a normal or "Gaussian" curve with the following characteristics:

Mean value:	$\mu = Np$	(1)
Sigma	$\sigma = \{(Np(1-p)\}^{1/2}$	(2)

Where:

N   Sample size (e.g., the number of coins tosses in each trial)
p   "Chosen" propensity
Np   Location of peak of distribution curve
σ "Sigma", a measure of the spread in the distribution

CHOSEN PROPENSITY p:

The concept of "chosen propensity" ("probability for success" or "chosen probability" in the Probability theory) has proven to be a potential source for confusion. "Binomial" means number 2. In binomial processes, even when there are many alternatives, they have to be narrowed to two: "Chosen" or "not chosen". For example, a die represents 6 individual possibilities. If we choose "1" then 2 to 6 will be "unchosen". If we choose "1 or 3" then 2, 4, 5, 6 are unchosen. Let's say we choose "1 or 3" then roll the die 10 times and get in succession: **1**, 5, **3**, 2, 6, **3**, 2, 5, **1**, 2. Since there are two 1's and two 3's, the count for our choice is 4 out of 10. All other numbers have no meaning because they were not chosen. (Had we chosen "1" then the count would be 2, because there are only two 1's.) How about choosing "1 and 3"? Answer: The count is guaranteed to be zero every time, because the choice requires two numbers to show up at the same time, but each time the die only gives one number.

MEAN VALUE (AVERAGE VALUE) $\mu = Np$

A distribution is an ideal graph, assuming that every thing works out as expected. Since the chosen propensity is p, if we try N times (sample size =N) we should get our choice Np times under this ideal condition. In reality most likely we will not get Np by trying the sample size N once. But if we repeat M times, each with the same sample size N, and take the average we should get close to Np. This is why Np is called the mean value.

SYMMETRY AND THE END OF PROBABILITY

THE BELL-SHAPED CURVE

Each point on the horizontal line corresponds to a possible count for our choice after trying a sample size N. For example if we choose heads and get 2 heads 8 tails after trying 10 coins (N=10), we get one count for the situation "2 heads 8 tails". The vertical position reflects the number of occurrences of a particular choice after many trials.

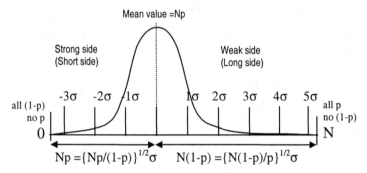

Figure 3: The normal approximation curve is the biased picture compiled from many trials by someone who has already "chosen" the event with propensity p as his or her measure for success. The sample size (of each trial) is N. If this event shows up in all samples (total success) the horizontal value is N, if it does not show up at all (total failure) the horizontal value is zero. The vertical value is the number of occurrences (trials) that have the same horizontal value. It is convenient to define a unit call sigma (σ) and use it to measure the horizontal length, so that instead of reading a long formula we can say the horizontal value is, say 2.5 sigma, or even more simply, 2.5 (then find out what sigma means later.)

Not surprisingly, the peak of the curve corresponds to the mean value. Horizontally, this peak is at the exact center when the chosen propensity p is equal to the unchosen propensity (1-p). For the general case, it is skewed toward the strong propensity, making this side of the curve shorter. The edge closest to the peak of the distribution curve corresponds to the case where the strong propensity (greater than 0.5) turns up 100% in the sample size N. It is therefore called the "strong side". The other edge is the "weak side". It corresponds to the case where the weak propensity (less than 0.5) turns up 100%. These last two situations are the limiting case, where the process meets the deterministic standard of classical physics.

SIGMA $\sigma = \{(Np(1-p)\}^{1/2}$

Sigma is a measure of the horizontal spread of a distribution. While sigma is a general concept, the sigma for binomial processes is of particular importance. This is because all distributions are related to the binomial distributions via the equal opportunity principle.

For mathematical convenience, it is customary to "normalize" sigma. This is done in several ways, depending on the application.

The first way is to divide sigma by the mean value. The goal is to express sigma as fractions of the mean to know, in relatively terms, how sigma and mean related to each other:

$$\sigma_M \equiv \sigma/(Np) = \{(1-p)/(Np)\}^{1/2} \tag{3}$$

The second way is to divide sigma by itself. The goal is to express everything in units of sigma's. This is convenient in analysis of the standardized normal curve. In this method of normalization, the horizontal coordinate will be read as 1, 2, 3... instead of 1 sigma, 2 sigma's, 3 sigma's (see also "number of sigma's" next.)

All normalized sigma's are called by the same name "sigma" for brevity. This may sound confusing, but should be clear from context.

NUMBER OF SIGMA'S

It is convenient to express the horizontal distance from the center of the distribution curve as a multiple of sigma's, and thus count horizontal lengths in "number of sigma's".

For the case $p < (1-p)$, the number of sigma's on the weak side of the curve is:

$$(m_\sigma)_W \equiv N(1-p)/\sigma = N(1-p)/\{Np(1-p)\}^{1/2} = \{N(1-p)/p\}^{1/2} \tag{4}$$

And for the strong side of the curve:

$$(m_\sigma)_S \equiv Np/\sigma = Np/\{Np(1-p)\}^{1/2} = \{Np/(1-p)\}^{1/2} \tag{5}$$

The symbol "$\equiv$" stands for "being defined as". The memory trick is to note that the weak side is longer, and thus has the higher number of sigma's. Alternatively we can also call the weak side as the long side, and the strong side as the short side, as long as we do not mix up p and (1-p).

B. SYMMETRY: THE MISSING PRINCIPLE OF DISTRIBUTIVE PROCESSES

The mystery of the average criterion

In chapter 2, we showed that the (probabilistic) proof of the Law of Large Number is internally inconsistent. By replacing the probabilistic version of the Chebyshev's inequality with its distribution version, we were able to establish an internally consistent non-probabilistic proof for the Law of Large Number.

In our investigation of the Law of Large Number, we encountered an unproven thesis that had been used in the probabilistic proof of the

Law of Large Number and was also instrumental in the construction of the new proof. This unproven thesis is the average criterion.

To refresh the reader's memory, the average criterion classifies series of equal length according to their average value. For example, in a series of 3 coin tosses we have 4 groups:

HHT, HTH, THH are all 2H1T, therefore identical.

TTH, THT, HTT are all 1H2T, therefore identical.

HHH stands alone.

TTT stands alone.

The average criterion is mind boggling because it is a global criterion. It implies, for example, that if we toss one coin at a time, the third toss is "aware" of the outcomes of the first two tosses. This "global awareness" seems counter-intuitive, but we must accept it because it is the only description that is compatible with reality.

The question is: Can we justify the average criterion with existing scientific knowledge? We will try to answer this question in the next several sections.

Why the Infinite Randomness Assumption must be wrong

To justify the average criterion we will return to the familiar process of tossing an unbiased coin repeatedly. We know from past experience that the long-term heads/tails ratio is 50/50, meaning that if we toss enough coins the counts for heads and tails will be approximately equal. For the moment the exact meanings of "enough coins" and "approximately equal" will not be our concerns (we will deal with them later in the book.) We will simply say that when the total tosses reach a certain number N, the count for heads N_H and the count for tails N_T will meet the "approximately equal" requirement. Note that this is an empirical fact which can be reproduced experimentally, not just a theory in the mind of the scientists.

Let's pretend that we have a record of a very long coin tossing experiment. First, we take N arbitrarily chosen consecutive tosses and call it series 1. Second, we divide series 1 into two sub-series, sub-series A consists of n_1 tosses, and sub-series B n_2 tosses. Third, we take the n_1 tosses following sub-series B and call it sub-series C. Finally, we examine series 2 formed by the sub-series B and C (also totaling N tosses) to see if heads and tails still meet the requirement of "approximately equal".

This experiment has been performed so many times in the history of statistics that it is not necessary to describe it in more detail. In fact, it is a standard procedure to verify the Law of Large Number. It suffices to say that if series 1 is "long enough", heads and tails will be approximately equal in both series 1 and 2, regardless how large or small n_1 is.

Let n_A and n_C be the number of heads in subsections A and C respectively. In order to maintain the condition "heads and tails are approximately equal in both series 1 and 2", it can be shown that the following must be satisfied:

$$n_A = n_C \pm \varepsilon \tag{1}$$

where ε is a number that can be made insignificant compared to both n_A and n_C. The reader may want to prove (1) as an intellectual exercise.

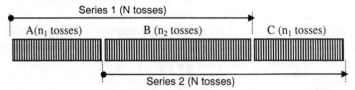

Figure 4: If N is large enough so that the counts for heads and tails meet the "approximately equal" requirement, it has been found experimentally that both series 1 and 2 (each with N tosses) would meet this requirement. This shows that distributive processes are independent of their starting point; which is possible only if there is a force maintaining a level of similarity between the two sub-series A and C.

It is clear that (1), which implies that n_C is numerically similar to n_A, is in conflict with the Infinite Randomness Assumption of the Probability theory, according to which n_C could range anywhere from 0 to n_1.

Since the Law of Large Number implies (1), we must conclude that the Infinite Randomness Assumption is wrong. Since this assumption is a founding principle of the Probability theory, we have confirmed once more that probability has no place in the Law of Large Number.

Distributive symmetry and distributive conservation
(Distributive processes and Noether's theorem)

For mathematical convenience, we will consider events 1, 2, 3, 4, etc. as points of a fictitious "event dimension". The equivalence of series 1 and 2 means that the heads/tails ratio of N consecutive tosses is not affected by any translational move in the event dimension. In other words, the long-term ratio is *conserved* with respect to the translation of events in distributive processes.

This conservation is possible because the n_1 points that are added (sub-series C) have approximately the same contribution to the long-term ratio as the n_1 points that are deleted (sub-series A) in a translation of n_1 events. By definition, a symmetry operation is one that does not cause any significant changes to the overall system. The translation of n_1 events in a distributive process is then a symmetry operation.

Readers who are familiar with Noether's theorem[2] know that for every conservation law there is a *continuous symmetry* and vice versa. Continuity per se does not exist in distributive processes, which are discrete by nature, but continuity still applies on an approximate basis to the average of many events. Thus, when enough events are considered together (as an ensemble), distributive processes should possess Noether-like symmetry and conservation. It is therefore not surprising that, in distributive process, long-term propensity (e.g., heads/tails ratio in the case of coin tossing) is conserved.

Readers who are familiar with recent scientific developments will recognize that long-term propensity is the "complexity" that emerges from the apparent randomness of individual events in distributive processes. Complexities can arise only if individual events are connected to one another by non-linear relationships. By assuming that the ensemble is the simple sum of its parts, classical probability is a linear model; therefore it must be wrong. (Frequency probability is unsatisfactory for other reasons.)

Symmetry, on the other hand, is a manifestation of order at the ensemble level. It is therefore the "force of non-randomness" that counters randomness to give rise to complexities. For differentiation purposes, we will refer to the symmetry and conservation exhibited by distributive processes as "distributive symmetry" and "distributive conservation" respectively.

Symmetry vs. Probability

We have just shown that distributive conservation is possible thanks to the existence of distributive symmetry. Since distributive conservation (e.g., the convergence of heads/tails ratio in a coin tossing process) is but a manifestation of the Law of Large Number, we must conclude that the Law of Large Number owes its root to symmetry. This is different from the position of the Probability theory, which claims that the Law of Large Number owes its root to probability. Since this claim by the Probability theory was refuted in chapter 2, we are almost certain that symmetry is the only valid root of the Law of Large Number. However, just to be sure, let's verify the difference between the two concepts one more time.

Since it is tricky to develop a general rule for symmetry, we will start with clear cut cases. Take the example of coin tossing with p=0.5 (i.e., long-term heads/tails ratio is 1-to-1.) Consider the following 5 sequences of N tosses each, where N is a very large number, say a billion:

Sequence 1: ... HTHTHTHTHTHTHTHTHTHTHTHTHTHTHT...
Sequence 2: ... HHTTHHTTHHTTHHTTHHTTHHTTHHTT...
Sequence 3: ... HHHHHHHHTTTTTTTHHHHHHHHTTTTTTT...
Sequence 4: ... HHHHHHHHHHHHHHHHHHHHHHHHHHHHHH...
Sequence 5: ... TTTTTTTTTTTTTTTTTTTTTTTTTTTTTT...

The most symmetrical sequence is 1 because it achieves symmetry (equal number of heads and tails) with every 2 tosses. Sequence 2 is slightly less symmetrical because it achieves symmetry with every 4 tosses. Sequence 3 is obviously more asymmetric than sequence 2. Sequence 4 and 5 are the most asymmetrical because they are blatantly biased toward heads and tails respectively. Thus, if symmetry is the criterion, the order of preference is 1, 2, 3, with 4 and 5 tied for last. Since N is very large, we are certain that neither 4 nor 5 can take place, because they are the most asymmetrical situations possible.

However, if probability is the criterion, all 5 sequences are equivalent because the probability of occurrence is $1/2^N$ for each. Thus, if probability is correct, sequences 4 and 5 are permissible.

The absence of long strings of exclusively heads (sequence 4) or exclusively tails (sequence 5), and the clear tendency of coin tossing process to lean toward a 1-to-1 heads/tails ratio in real experiments then force on us the decision to pick symmetry over probability.

We can now safely announce that (distributive) symmetry, and not probability, is the governing rule of distributive processes.

The symmetry justification for the Law of Large Number

Let's go into more detail on how symmetry operates in distributive processes. As far as symmetry is concerned, the four 3H1T combinations HHHT, HHTH, HTHH, and THHH are "symmetrically equivalent" because –by setting H=1 and T=0 for convenience- the average value for each one of these combinations is 3/4. HHHH is different because its average value is 4/4=1. For this reason, the four 3H1T combinations are treated as one group, separately from HHHH.

As we have seen in chapter 2, this is exactly the logic of the average criterion, which leads to the Law of Large Number. We thus conclude that symmetry is the correct justification for the average criterion.

The equal opportunity principle
as the compromise between symmetry and randomness

We can now construct the working mechanism of distributive processes. Randomness allows multiple combinations, instead of a single combination, to occur; and symmetry keeps the maximum deviation of each combination under control by recognizing the differences and similarities among all combinations. Since randomness and symmetry are intertwined, we cannot observe them separately. What we can observe is their combined effect, which we have referred to in the first chapter as "the equal opportunity principle".

The equal opportunity principle, then, is not the work of randomness alone as incorrectly assumed by the Probability theory, but a compromise of randomness and symmetry; and the relative strengths of these two forces decide how random or well-controlled a process will be.

For the particular case of coin tossing, the rules are as follows:

1. Symmetry considers combinations with the same Heads-Tails discrepancy as belonging to one group. It combines the sum of individual propensities to determine whether a group is prohibited or allowed to occur.

2. The presentation of a combination in a distribution depends on its combined propensity and the relative strength of randomness and symmetry.

3. Under the restriction imposed as a compromise between randomness and symmetry, events with propensities below the prohibition threshold cannot occur. *(Note: This is only approximately correct. We will add the correction later.)*

4. Also under the restrictions imposed as a compromise between randomness and symmetry, events with propensities above the prohibition threshold can only vary within certain limits. *(Note: Same as 3 above.)*

Obviously these rules will not hold in processes of infinite randomness. We therefore must emphasize once more that the Infinite Randomness Assumption of the Probability theory is necessarily wrong. It is the "happy middle" between randomness and non-randomness that gives birth to distributive processes.

Maximum asymmetry and correction to the Law of Large Number

Earlier we have ruled out an infinite string of consecutive heads or tails because it represents extreme asymmetry and therefore violates the principle of symmetry. We also found that symmetry cannot be isolated from randomness. Thus, the symmetry that we can observe in distributive processes is the residual symmetry, i.e. the net symmetry after randomness has been taken into account.

Understandably, the net symmetry can vary from process to process. This means the maximum number of consecutive heads N_{max} may differ in two processes with the same propensity p. We call N_{max} the maximum asymmetry of the process.

Let's consider the following sequences that theoretically could take place in the tossing process of an unbiased coin with sample size N=20:

1. HHHHHHHHHHHTTTTTTTTTT
2. HHHHHHHHHHTTTTTTTTTTHT
3. HHHHHHHHTTTTTTTTTHHTT
4. HHHHHHHTTTTTTTTHHHTTT
5. HHHHHHTTTTTTHHHHTTTT

Since each combination consists of 10 heads and 10 tails, the average value for each is identical and equal to 10/20=1/2. According to existing knowledge of the Law of Large Number, these five combinations are equivalent and belong to the group that has the following probability:

$$20!/\{10! \times 10!\}/2^{20} = 0.176$$

We note, however, that sequence 1 has 10 heads in a row and 10 tails in a row, sequence 2 has 9 in a row, sequence 3 has 8 in a row, sequence 4 has 7 in a row, sequence 5 has 6 in a row.

Now consider a particular coin tossing process:

If $N_{max} \geq 10$:   All 5 sequences can be realized.

If $N_{max} = 9$:   Sequence 1 cannot be realized

If $N_{max} = 8$:   Sequences 1 and 2 cannot be realized

If $N_{max} = 7$:   Sequences 1 through 3 cannot be realized

If $N_{max} = 6$:   Sequences 1 through 4 cannot be realized

If $N_{max} \leq 5$:   Sequences 1 through 5 cannot be realized

The combinations that cannot be realized are called "prohibited combinations". Since the Law of Large Number is based on the assumption that there are no prohibited combinations, it will have to be corrected accordingly. For example, if $N_{max} = 9$ there will be 20 prohibited 10H10T combinations, each with 10 consecutive heads or tails. They are:

1. HHHHHHHHHHHTTTTTTTTTT
2. THHHHHHHHHHHTTTTTTTTT
3. TTHHHHHHHHHHHTTTTTTTT
4. TTTHHHHHHHHHHHTTTTTTT
5. TTTTHHHHHHHHHHHTTTTTT
6. TTTTTHHHHHHHHHHHTTTTT
7. TTTTTTHHHHHHHHHHHTTTT
8. TTTTTTTHHHHHHHHHHHTTT
9. TTTTTTTTHHHHHHHHHHHTT
10. TTTTTTTTTHHHHHHHHHHHT
11. TTTTTTTTTTHHHHHHHHHHH
12. HTTTTTTTTTTTHHHHHHHHH
13. HHTTTTTTTTTTTHHHHHHHH
14. HHHTTTTTTTTTTTHHHHHHH
15. HHHHTTTTTTTTTTTHHHHHH
16. HHHHHTTTTTTTTTTTHHHHH

17. HHHHHHTTTTTTTTTTHHHH
18. HHHHHHHTTTTTTTTTTHHH
19. HHHHHHHHTTTTTTTTTTHH
20. HHHHHHHHHTTTTTTTTTTH

The mathematics for this kind of correction is understandably cumbersome, but straightforward. We will not attempt to perform it here (although the reader is more than encouraged to work out at least one case to convince him or herself that we are dealing with more than a cosmetic fix of the Probability theory.)

The main point is that, under certain circumstances, the corrected Law of Large Number could be quite different from the one given in existing textbooks on probability or/and statistics. This point is important because the main justification for the Probability theory is that it gives the same results as the textbook formula for the Law of Large Number. In cases where the Law of Large Number needs to be corrected, the probability predictions will be wrong!!!

Thus, reality again forces on us the conclusion that the underlying principle of distributive processes is symmetry, not probability.

Distributive symmetry and system memory

In a coin tossing process, the probability theory takes for granted that each toss has nothing to do with any other toss or tosses. Since we have chosen symmetry over probability, this point has to be discarded. This means at least some tosses must be dependent on other tosses. Moreover, since tosses follow one another, the general implication of symmetry is that future events are not independent of past events, but are related to them. In other words, distributive processes have memory!!!

This result is definitely controversial. In fact, it is more than controversial. To the proponents of the Probability theory, it is the ultimate nonsense! But is it really nonsense? Think about this: The idea of process memory is controversial to us and sounds non-sensical to probability proponents only because we have been conditioned to accept the indifference principle of probability as truth. We have forgotten that the indifference principle is no more than a belief started by the founders of classical probability and perpetuated by probability proponents of later generations. The ultimate validity test of a scientific principle is empirical result. Thus, the disagreement between the indifference principle and empirical result does not allow us to conclude that the empirical result is non-sensical. We must instead take the view that the indifference principle is wrong and accept process memory as a curious reality forced on us by Nature.

Process memory brings up the question of time. Is it possible that the nature of time has been completely misunderstood in prior analyses of distributive processes? We will return to this puzzling question in a dedicated discussion following chapter 7 "symmetry, synchronicity, and the meaning of space-time".

C. SYMMETRY SOLUTION TO THE LONG-RUN PARADOX

The long-run paradox seems to have been successfully solved when we determined that the longest string of heads (or tails) must be of finite length in a coin tossing process to meet the requirement of symmetry. Note, however, that we have treated coin tossing processes as if they were governed only by randomness and symmetry. This is not completely true because, as we have pointed out in chapter 1, at least some of the outcomes in a series of tosses may be predicted with the deterministic laws of Newtonian mechanics. We will see that this added element of determinism further weakens the probability hypothesis and makes the case for symmetry even stronger.

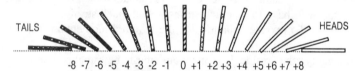

TAILS   HEADS

-8 -7 -6 -5 -4 -3 -2 -1   0 +1 +2 +3 +4 +5 +6 +7 +8

Figure 5: Possible orientations of a coin when it
makes the last contact with the settling surface.

Figure 5 shows the many possible situations that a tossed coin may experience when it makes the last contact with the horizontal surface available to it. By "last contact" we mean the last touch when the coin might still have a chance to defy the simple downward pull of gravity on its mass. There should be no conceptual problem here, because this "last contact" does exist for every toss.

By symmetry considerations, there is no reason to rule out any of the 17 possibilities illustrated in figure 5, as well as other possibilities in between, when a great many tosses are attempted by, say, a human being.

Given that there may still be some residual momentum in the coin at this final stage of its motion, we have the following judgments:

TABLE 1: Judgments on coin toss outcome

Situations	Judgment of outcome
+5 to +8	Certainly landing on heads
+3 and +4	Leaning strongly toward heads
+1 and +2	Leaning toward heads

0	Undecided
-1 and -2	Leaning toward tails
-3 and -4	Leaning strongly toward tails
-5 to -8	Certainly landing on tails

We could therefore conclude that when a coin makes it last contact with the settling surface, its tendency toward heads or tails varies from complete uncertainty (situation 0) to complete certainty (situations -5 to -8 and +5 to +8); and if we toss a great many coins, all of these tendencies will be represented.

We will now re-examine the residual (spinning) momentum in more detail. When we say that situation +5 certainly leads to heads, we have already made a subjective judgment on the magnitude of the residual momentum. It is entirely possible that the residual momentum is much stronger than our expectation. We therefore have to allow the possibility that –despite our belief that every coin in situation +5 will land on heads- this unusually strong momentum would push the coin in the counter clockwise direction and force it to land on tails. However, to produce the same reverse effect in situation +6, the residual momentum will have to be much stronger, and much stronger still with situation +7. To reverse situation +8, the required momentum must be extremely extremely large.

In reality the residual momentum must be smaller than the "maximum momentum", which is the sum of the initial momentum that the tosser imparts on the coin plus external effects. Thus, the residual momentum will range anywhere from being very close to zero to a value smaller than the maximum momentum, which is itself finite. We can safely say, therefore, that even if the spinning momentum of the coin always conspires against heads; it still cannot prevent heads from occurring.

But there is no reason for us to assume that the spinning momentum always conspires against heads. By symmetry we must assume that there exist also cases where it would reinforce heads. With this in mind, we are now certain that if we throw enough tosses, there must exist a non-zero number of outcomes that are heads. Furthermore, the larger the number of tosses, the more heads will be in the finally tally!

Referring back to table 1, in order to have any hope at all that heads will not show up somewhere in the long run, we must at least eliminate situations +5 to +8. There is no doubt that this is achievable (by tossing the coin with a modern robot arm programmed for the task, for example.) However, such a manipulation would disqualify the process from being a proper coin tossing process. We are forced to conclude, therefore, that in a proper coin tossing process, heads will certainly show up in the long run. By applying a similar argument, tails also certainly will show up in the long run.

But could a long run mean an infinite number of tosses? The answer is again "No!" because this would create infinite asymmetry, a blatant violation of symmetry. We conclude therefore that the longest strings for consecutive heads and consecutive tails must be finite.

This complete our solution to the long-run paradox that we posed in chapter 1. The success of our analysis further confirms that symmetry is the governing law of distributive processes.

D. SYMMETRY SOLUTION TO THE GIBBS PARADOX

In the last chapter we have pointed out the conflict between probability and the solution to the Gibbs paradox. To refresh the reader's memory, the Gibbs paradox concerns the change in entropy when two compartments containing the same gas under the same condition are connected to each other. Since is no increase in randomness of the total system, the entropy change should be zero ($\Delta S=0$), but the probability theory predicts wrongly that there is an increase in entropy ($\Delta S>0$).

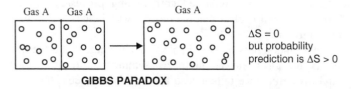

GIBBS PARADOX

Figure 6: The mixing of identical gas should cause no entropy increase, but the probability prediction is $\Delta S >0$. This is the Gibbs paradox. The Gibbs paradox is removed by recognizing that the underlying principle is symmetry. In other words, there is no need for the concept of probability in this process.

Historically, it was Josiah Willard Gibbs who first recognized the paradox, which explains why it bears his name. Gibbs also noticed that the paradox could be solved by considering all gas atoms together as an ensemble. From this ensemble standpoint it would make no difference, for example, if atom X is in state 1 and atom Y in state 2, or atom Y is in state 1 and atom X in state 2. It naturally follows that the unmixed state (left picture of figure 6) and the mix state (right) have the same measure of randomness, therefore there is no increase in entropy.

Gibbs' solution was considered *ad hoc* when it was proposed because the statistical theory that he corrected was based on the Probability theory. In the probability approach, one uses the properties of the parts (the individual atoms) to deduce the properties of the whole (the ensemble of atoms). In this formalism, it would not make sense for the effect (the ensemble) to affect the cause (the individual atoms); but that was exactly Gibbs' solution was all about.

Gibbs's solution was finally confirmed to be valid by a logic credited to quantum mechanics. What is this logic? The alert reader must have guessed it correctly: SYMMETRY! Of course!

Yes! It was system symmetry that finally demystified the Gibbs paradox. Thus, the Gibbs paradox is one of the most powerful proofs that symmetry, and not probability, is the underlying principle of distributive processes!

E. SYMMETRY AND THE END OF PROBABILITY

The remaining question at this point is: "Can the Probability theory be corrected by taking symmetry into account?" We will use the familiar process of coin tossing to argue that this cannot be done.

N_{max} is a measure of the maximum asymmetry allowed by the process before symmetry makes the necessary correction. For this reason it is a convenient measure of the force of randomness, which gives rise to asymmetry. Since there is absolutely no reason for N_{max} to be the same for all coin tossing processes, it is more likely than not that two arbitrarily chosen unbiased coin tossing processes will have 2 different N_{max}'s. It is clear that the process with higher N_{max} is the more random process.

To put randomness in measurable terms, we note that if the first toss is heads, the ideal way to minimize asymmetry is to get tails in the second toss. But since tails is also a state of asymmetry, the third toss will have to be heads, then the fourth toss will have to be tails, etc. The result is a repeating series of heads and tails. Since the only unknown is the first toss, the level of randomness for this process must be at the lowest value possible. For mathematical convenience, we will say that this is a process with zero randomness!

Understandably the maximum string of heads or tails for a process of zero randomness must be at its smallest possible (1 for the case of unbiased coin tossing.) The stronger the force of randomness is, the longer the length of the maximum strings should be. When the maximum string is allowed to approach infinity we have a process of infinite randomness!

Mathematically then, the range for process randomness is between zero and infinity. Since infinite randomness violates symmetry, we only have two options. The first option is to accept that all distributive processes are governed by infinite randomness. The second option is to accept that all distributive processes must obey symmetry and therefore can only have finite randomness. Our choice is obvious: Symmetry and finite randomness.

We already know that infinite randomness is the built-in assumption of classical probability (i.e., the Infinite Randomness Assumption). In order to incorporate symmetry, the Infinite Randomness Assumption will have to be discarded. The problem is, this action would make it impossible for all equivalent events to have the same probability, which is the required starting point of classical probability. Thus, classical probability cannot be saved.

The Popperian propensity interpretation of probability suffers the same fate as classical probability, because it assumes that each equivalent event in a process (i.e., events occurring under the same experimental condition) has the same propensity. Thus, the (Popperian) propensity interpretation also cannot be saved.

Symmetry can be added to frequency probability by modifying the convergence axiom (to take into account of events that will be absent in the long-run) and abandoning the randomness axiom. We should not forget, however, that frequency probability relies on classical probability for its calculations of long-term ratios and therefore must assume infinite randomness. With classical probability out of the picture and the only thing left is a guarantee of long-term convergence by virtue of symmetry, is there any reason to retain the label "probability" in the name of the theory? The answer is NO. Thus, the frequency Probability theory -as we know it- also cannot be saved.

But do we really need to save the Probability theory? Historically, probability was added to science as an attempted solution to problems that could not be handled by deterministic methods. We now know that these problems can be handled successfully by symmetry. Since symmetry has always been a well accepted scientific principle, its new role in distributive processes makes the Probability theory unnecessary.

Thus, with the emergence of symmetry we have finally converged to a reasonable description of distributive processes; and we have also reached THE END OF PROBABILITY[3]!

First written November 2002, revised December 2002

NOTES:

1. Reference: "Propensity, Probabilities, and the Quantum Theory", Karl Popper, 1957; reprinted in "Popper Selections", edited by Miller, Princeton University Press, 1985.

The following quote from this document explains Popper's meaning of the term "propensity".

"We thus arrive at the propensity interpretation of probability. It differs from the purely statistical or frequency interpretation only in this —that it considers the probability as a characteristic property of the experimental arrangement rather than the property of the sequence.

"The main point of this change is that we now take as fundamental the probability of the result of a single experiment, with respect to its conditions, rather than the frequency of results in a sequence of experiments."

While the meaning of the second paragraph is somewhat vague, it appears that Popper meant to apply the concept of propensity to each individual event. This was unfortunate because this would lead to a dead end similar to that of classical probability. It was also very regrettable, because Karl Popper was so close to a more complete concept of distributive process compared to classical probability and frequency probability.

2. Noether's theorem basically states that "For every continuous symmetry there exists a conservation law and vice versa".

3. Note that subjective probability, which allows individual probability to vary from case to case, is not affected by the elimination of the Probability theory.

APPENDIX
MORE ON THE CONCEPT OF PROPENSITY
The replacement of probability by propensity

The purpose of this appendix is to give a deeper account for the term "propensity" that we introduced at the beginning of this chapter as the Distribution theory's counterpart for the term "probability" in the Probability theory, which we believe should be discarded. The concept of "propensity" was borrowed from Popper, but has been modified to fit the Distribution theory. The Popperian meaning of propensity is a tendency present in every event, whereas our meaning of propensity is a long-term tendency exhibited by long enough series of many events.

If we flip a coin repeatedly, the number of occurrence N_H for heads will converge -within certain tolerance- to the following ratio:

$N_H/N = p$ (1a)

Where N is the number of total occurrences, including both heads and tails.

Classical probability says that the "event probability" for heads to occur is p, meaning that the chance for heads to occur is fixed at p for every single trial. This probability is of course unverifiable because heads either occurs or it doesn't.

By examining the situation carefully, we are forced to conclude that the concept of fixed probability was the result of an unfortunate confusion. Classical probability has mixed up the parts with the whole. The parts are the single events of individual coin flips, the whole is the final distribution, which is formed by the process of convergence of many parts.

In hindsight, this confusion may have been caused by a wrong interpretation of the mathematical procedure used by classical probability in its development stage.

Symmetry and the end of probability

We now know that the convergence mechanism in distributive processes is an averaging process. Thus, p only has meaning of an average tendency manifested by many tosses.

The average property p of many tosses can be considered mathematically as property p of an "average toss". Clearly, this "average toss" is not a real toss because it is just a fictitious artifact invented as a part of the calculation procedure for mathematical convenience.

Classical probability obviously has mistaken this fictitious "average toss" for a real toss. Once this mistake had been made it was only natural to connect p, which is a property of the fictitious average toss, to a real toss; and we ended up with the idea of fixed event probability that has lasted three hundred years.

The "fixed event probability" mistake taught us an important lesson. Extreme care has to be exercised while using mathematics to solve a physical problems, or important physical meanings may be lost or misunderstood amid the jungle of mathematical symbols and equations.

To eliminate confusion, we will use the term "inherent propensity" or simply "propensity" instead of "probability" or "event probability" in all discussions from here on forward. It is important to define the exact meaning of "propensity".

"Propensity" is the average property of a converging distribution. Since distributions are difficult to visualize, it is customary to identify propensity as the property of a fictitious "average event" representative of the distribution. "Average event" is a mathematical concept invented for convenience only and has nothing to do with actual events.

The philosophical implications of propensity

We will start with an arbitrary distribution A whose possible single event values are confined in the range (a, b) and whose average value is μ_A. Obviously if only a single event is considered, we will find that its value X_A satisfies:

$$a \leq X_A \leq b \tag{2a}$$

which is not very interesting. However if we take N samples of distribution A at a time, calculate their average value and use it as a point in a fictitious distribution B, i.e.

$$X_B = (\Sigma X_{Ai})/N \qquad i=1 \text{ to } N \tag{3a}$$

we will find that X_B also satisfies:

$$a \leq X_B \leq b \tag{4a}$$

Moreover, as we have learned from the proof for the Law of Large Number in the appendix of chapter 2, the average value of distribution B is the same as that of distribution A:

$$\mu_B = \mu_A \tag{5a}$$

It is the similarities exhibited in (4a) and (5a) between B and A that makes B an interesting distribution.

Let's take a look back at (4a). Although this relationship holds for all N's, when N increases to certain point N_a, symmetry requirement will make it impossible for all N samples to have value a; also when N increases to certain point N_b, it will be impossible for all N samples to have value b. As a result, X_B obeys the following relationship:

$$a_N \leq X_B \leq b_N \tag{6a}$$

Where a_N and b_N are monotonous functions of N, and the range (a_N, b_N) keeps decreasing when N increases. When N approaches infinity, this range approaches zero. At this limit (6a) becomes:

$$. \; a_N = X_B = b_N = \mu_A \tag{7a}$$

We notice that result (7a), which came from a symmetry argument, agrees with the Law of Large Number. Since the Law of Large Number is in conflict with the theory of Probability (Appendix of chapter 2), this is another supporting evidence for symmetry over probability.

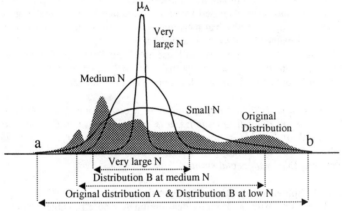

Figure 1a: Development of distribution B, formed by taking the average of N samples from original distribution A. Note that at very large N, the range of values for B is decreased, and B shows a clear propensity toward the average value μ_A of the original distribution A.

When N approaches infinity, the width of distribution B approaches zero and the propensity becomes identically μ_A.

We can take the mean value μ_A as the propensity of the original distribution A. Why are we allowed to do so? Because if there is no propensity, there should be only uncontrollable randomness, not a well defined distribution.

If we examine A by taking one sample at a time, it would be impossible to determine its propensity μ_A. By increasing the sample size N, the propensity μ_A becomes more well defined but some degree of uncertainty still exists. This degree of uncertainty decreases with N, and disappears when N approaches infinity. From this reasoning, it is clear that the determination of propensity requires multiple samples.

(We will learn in later chapters that when N is large enough, the set of X_B's forms an approximately normal distribution, although this knowledge is not necessary for the present discussion).

Chapter 4

Symmetry
and the Law of Large Number

We will show that symmetry naturally leads to the Law of Average, which turns out to be a more precise version of the (existing) Law of Large Number. The success of symmetry confirms that the Infinite Randomness Assumption (inherent with the Probability theory) is incorrect; and the randomness level must be finite for all distributive processes.

For this reason we will replace the Probability theory (which is based on infinite randomness) with the Distribution theory (which is based on finite randomness).

A. SYMMETRY LOGIC FOR THE LAW OF AVERAGE

The symmetry approach to distributive process

The possible orientations of a tossed coin can be divided into three categories, as seen in figure 1.

A. Undecided (could end up heads, could end up tails)
B. Definitely tails.
C. Definitely heads.

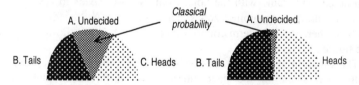

Figure 1: The three regions of a coin toss. Note that, if at all, classical Probability theory is only applicable to a very small part of region A. Since this region could vary from case to case, classical probability cannot be the correct description of real life distributive processes.

It can be seen that the Probability theory assumes that the only region that exists is the "undecided region" A. While this is mathematically possible, from our discussions in earlier chapters we must assume that region B (definitely tails) and C (definitely heads) also exist in the general case. This shows clearly that if the Probability theory could be applied at all, its scope would be limited to a very extreme class of phenomena.

Let's perform a thought experiment. In this experiment we toss a great number of N coins at a time for a great many times M so that all possible combinations are present in the data, then use each trial (of N samples) as a data point to form a new distribution. Mathematically the number of heads N_H in an arbitrarily chosen trial could range any where from 0 to N; but because there exist cases of "definitely heads" and "definitely tails", if N is large enough the smallest possible number of heads in a trial N_{min} will be larger than 0; likewise the largest possible number of heads in a trial N_{max} will be smaller than N.

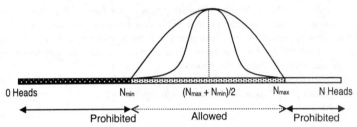

Figure 2: Although we still do not know much about the allowed zone, the Principle of Symmetry tells us that it must be symmetrical across the center. We also know that the minimum density has to be at the edge, and the maximum at the center. This gives the curve the shape of a bell. Two possible bell-shaped curves are shown in the picture.

If we plot the number of occurrences versus the count for heads, we will see three zones. The zone from 0 heads to $(N_{min}-1)$ heads must have zero count. The same with the zone from $(N_{max}+1)$ heads to N heads. These are the two prohibited zones. The zone from N_{min} heads to N_{max} heads is where the long term counts are non-zero. It is the allowed zone (see figure 2).

The situation in the allowed zone is fuzzy. However, symmetry tells us that the shape of the distribution curve in this zone must be symmetrical across the center. Also the two edges of the allowed zone must correspond to the minimum counts, while its center must correspond to the maximum counts. Two curves that satisfy these conditions are shown in figure 2. They can be recognized as the "bell-shaped" curves of binomial distributions.

This concludes our qualitative argument for a new approach for distributive process. We will now proceed to quantify this approach.

Binomial average

All of us are familiar with the averaging operation c= (a+b)/2. For example if a=3, b=6, then their average is c= (3+6)/2=9/2=4.5. For further reference, we will call a result of this averaging process the "arithmetic average" or "numerical average".

The (numerical) averaging process is so simple and natural that we tend to forget that, in order for it to have meaning, both a and b have to exist in the same environment.

Now think of the case where the existence of a and b are mutually exclusive of each other; meaning that if a exists then b does not exist, and vice versa. This is not just a mathematical possibility, because a and b do represent two opposite deviations from the long term average in distributive processes.

Obviously a and b are different, but at the same time they have to be, in some sense, equivalent. Can we then say that their average is zero? No, we can't. Think of two persons. The first person switches between owning and owing a million dollars, the second person between owning and owing a hundred dollars. The numerical average is zero for both, but obviously the first person has a more "dynamic" life than the second.

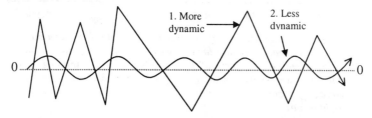

Figure 3: The numerical averages for cases 1 and 2 are both zero. However, it is clear that case 1 has more variations than case 2. This shows that numerical average is not an adequate description of distributive processes.

This example forces the need for a new concept of average to measure the dynamic of fluctuations: The average of mutually exclusive potentials! We will call this the "binomial average". Since symmetry operates on negative and positive deviations, which are mutually exclusive of each other, it is clear that its rule is "binomial average".

In our search for a method to calculate the binomial average, we need to return to arithmetic.

Consider two points a and b in a rectangular coordinate system, with a confined to the horizontal direction, and b the vertical direction. If a is different from 0 (i.e., a exists) then we are certain that b=0 (i.e., b does not exist), and vice versa. Thus, a and b are perfect mathematical representatives of exclusive potentials.

Geometrically, the multiplication operation is a doubling of dimension (one-dimensional line segments transformed into two-dimensional area). The square root of the product of a and b (i.e.,

$c=[ab]^{1/2}$) is the side of the square that has the same area as their product ab. In terms of dimensions, the operation is as follows: Double the dimension of a and b, then reduce the dimension by half to bring them back to their original dimension, with all differences equalized.

(a, b) => ab => (c, c)

The net effect is:

(a, b) => (c, c) where $c = [ab]^{1/2}$

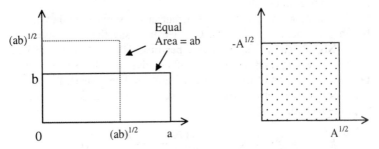

Figure 4: The square root of the product of two number a and b (i.e., $[ab]^{1/2}$) is a good measure for the average behavior of two mutually exclusive entities (left picture). On the other hand, the splitting of an entity A to two equivalent but mutually exclusive entities is a binomial averaging process, which gives $A^{1/2}$ to each of the two new entities (right picture)

The process of transforming two numbers to a new number that represents both equally is of course the averaging process; and since symmetry treats positive and negative deviations equally, this averaging process meets the requirement of symmetry that we are looking for. The (symmetrical) binomial average of two mutually exclusive entities a and b is therefore $[ab]^{1/2}$. Readers who are familiar with arithmetic should recognize this as the "geometrical average" of a and b.

What about the case an entity A splits itself into two equivalent but mutually exclusive potentials? The answer is: The reverse process applies. The binomial average resulting from the splitting of A is therefore $A^{1/2}$. Since the exclusive potentials act on the reference from different directions, their mathematical forms are $+A^{1/2}$ and $-A^{1/2}$.

Now we are ready to return to the binomial process of averaging N samples.

Symmetry logic for the Law of Average

We will start with an arbitrary "original distribution" that possesses an average value p. We will take N samples from this distribution at a time, then use their average value to form a new distribution. Since it is trivial to show that the average value of the new distribution is also p, we will simply use this result without offering a proof for it.

There is no loss of generality by considering only the case $0 \leq p \leq 1$. Since p is an overall value applicable only to all N samples, not any particular individual sample, there is no reason to appeal to the concept of probability. We will instead consider p as a parameter that decides the long term tendency of the averaging process under investigation. As agreed earlier, we will call p the propensity of the process.

This formalism allows us to consider the averaging process of N samples as a binomial distribution where p is the propensity for success.

For a trial of sample N, the actual count could be either positive or negative relative to the mean value. The set of positive discrepancies could take any integer value from 0 to +N. Likewise the set of negative discrepancies could take any integer value from 0 to –N. Since positive and negative values are mathematically equivalent, the discrepancies form two equivalent sets of numbers, each of exactly the same size as N. On the other hand, they are mutually exclusive of each other because if a deviation is positive then it is impossible for it to be negative at the same time, the same logic applies to a negative deviation.

We will now appeal to symmetry: The positive and negative deviations must be equivalent to each other, because there is no reason for them not to be so. This allows us to argue that the potential deviations that arise from N are mathematically $+N^{1/2}$ and $-N^{1/2}$ (in positive direction and negative direction away from the mean value, respectively).

There are two more factors that have to be taken into account. First, the sampling process is itself an unknown factor. To achieve the same level of accuracy, a larger sample size is required for a more random (or equivalently, less symmetrical) sampling process. Second, on the average, for p occurrences of the propensity for success, there will be (1-p) non-occurrences. Occurrences and non-occurrences are also mutually exclusive of each other. By appealing again to symmetry, both factors can be handled by taking the average of kp and k(1-p), where k is a non-negative process-related parameter, giving $k\{p(1-p)\}^{1/2}$; which is applicable to both positive and negative deviations.

The final results are:

Positive deviations of averaging process:	$+k\{Np(1-p)\}^{1/2}$	(1)
Negative deviations of averaging process:	$-k\{Np(1-p)\}^{1/2}$	(2)

The term $\{Np(1-p)\}^{1/2}$ is known as the standard deviation sigma (σ) for binomial processes. It is well known that binomial distributions can be approximated by a normal distribution when N is large enough.

Let p be the propensity for success, N_S the actual number of successes in any given trial. It follows from (1) and (2) that:

$$Np - k\{Np(1-p)\}^{1/2} \leq N_S \leq Np + k\{Np(1-p)\}^{1/2} \qquad (3)$$

This leads to:

$$p - k\{p(1-p)/N\}^{1/2} \leq N_S/N \leq p + k\{p(1-p)/N\}^{1/2} \qquad (4)$$

It can be seen from (4) that, if k is finite, when N approaches infinity N_S/N approaches p. This means the average value of N samples converges; which completes our proof for the Law of Average.

Note that in our derivation of the Law of Average we did not use the concept of probability at all.

B. SYMMETRY AND THE LAW OF LARGE NUMBER

Symmetry vs. Probability in long-term distributions

Let's investigate a distributive process with low sample size, say the tossing of 4 unbiased coins. The mathematically possible combinations are:

TABLE 1: Result for longest string ≥ 4

Case	Detail	Possible combinations	Count
1	4H0T	HHHH	1
2	3H1T	HHHT, HTHH, HHTH, THHH	4
3	2H2T	HHTT, HTTH, HTHT, TTHH, THHT, THTH	6
4	1H3T	TTTH, THTT, TTHH, HTTT	4
5	0H4T	TTTT	1
OVERALL			16

The main success of the Probability theory is it ability to predict long term ratios based on analyses such as that shown in table 2. We should note, however, that this particular analysis works only because most distributive processes have no problem in producing strings of 4 heads and 4 tails. Needless to say that in such a case symmetry –via the binomial formula- predicts exactly the same ratios, namely:

TABLE 1a: Long-term ratio predictions

4H:	1/16
3H1T:	1/4
2H2T:	3/8
1H3T:	1/4
4T:	1/16

Let's now consider a hypothetical process with very low randomness, such that the longest strings allowed are 3 heads and 3 tails. Table 1 would have to be modified as followed:

TABLE 2: Result for longest string = 3

Case	Detail	Allowed combinations	Count
1	4H0T	~~HHHH~~	0
2	3H1T	HHHT, HTHH, HHTH, THHH	4
3	2H2T	HHTT, HTTH, HTHT, TTHH, THHT, THTH	6
4	1H3T	TTTH, THTT, TTHH, HTTT	4
5	0H4T	~~TTTT~~	0
OVERALL			14

For the case of table 2, symmetry will successfully predict the absence of 4H and 4T, giving the following long term ratios:

TABLE 2a: Long-term ratio predictions by Symmetry

4H:	0
3H1T:	2/7
2H2T:	3/7
1H3T:	2/7
4T:	0

The Probability theory, on the other hand, must stay with table 1a and would make the wrong predictions.

The hypothetical cases with longest string = 2 and longest string = 1 are listed below to amplify the point that we will make shortly.

TABLE 3: Result for longest string = 2

Case	Detail	Allowed combinations	Count
1	4H0T	~~HHHH~~	0
2	3H1T	~~HHHT~~, HTHH, HHTH, ~~THHH~~	2
3	2H2T	HHTT, HTTH, HTHT, TTHH, THHT, THTH	6
4	1H3T	~~TTTH~~, THTT, TTHH, ~~HTTT~~	2
5	0H4T	~~TTTT~~	0
OVERALL			10

TABLE 3a: Long-term ratio predictions by Symmetry

4H:	0
3H1T:	1/5
2H2T:	3/5
1H3T:	1/5
4T:	0

TABLE 4: Result for longest string = 1

Case	Detail	Allowed combinations	Count
1	4H0T	~~HHHH~~	0
2	3H1T	~~HHHT, HTHH, HHTH, THHH~~	0
3	2H2T	~~HHTT, HTTH,~~ HTHT, ~~TTHH, THHT,~~ THTH	2
4	1H3T	~~TTTH, THTT, TTHT, HTTT~~	0
5	0H4T	~~TTTT~~	0
OVERALL			2

TABLE 4a: Long-term ratio predictions by Symmetry

4H:	0
3H1T:	0
2H2T:	1
1H3T:	0
4T:	0

The low-randomness processes in tables 2 through 4 were given as examples to demonstrate the conceptual superiority of symmetry over probability. But since processes with such low-randomness are not typical, usually for the range of sample size N of interest probability and symmetry have equal predictive power.

The situation changes significantly at high sample size. To see this we will look again at the set of computer data reported in chapter 1, where 263 strings of 29 heads were predicted by the Probability theory, and none recorded in 141.3 billion tosses.

If we consider 30 tosses as one event it is clear that the following combinations cannot occur:

30 consecutive heads

30 consecutive tails

29 consecutive heads followed by 1 tails

29 consecutive tails followed by 1 heads

1 tails followed by 29 consecutive heads

1 heads followed by 29 consecutive tails

These prohibited combinations are also in disagreement with the Probability theory, which predicts that all of them should occur in the long run.

We reach two conclusions:

First, the success of the Probability theory in predicting long-term ratios has been a case of lucky coincidence. This lucky coincidence was repeated time and time again because the weakness of the Probability theory is not visible in processes with low sample sizes. It is only at high sample sizes that the weakness of the Probability theory is finally exposed.

Second, symmetry succeeds not only where probability succeeds (at low sample sizes), but also where probability fails (at high sample sizes). It is therefore the correct foundation of distributive processes.

TABLE 6: Computer simulation of 141.3 billion coin tosses
(by Hanspeter Bleuler, Germany, private communication, November 2002)
Counts are combined total for strings that are all-heads

STRING	PROBABILITY PREDICTION	ACTUAL COUNT
1	70,652,392,105	70,650,390,138
2	35,326,196,053	35,326,178,115
3	17,663,098,026	17,663,595,966
4	8,831,549,013	8,832,056,857
5	4,415,774,507	4,416,137,049
6	2,207,887,253	2,208,074,053
7	1,103,943,627	1,104,163,729
8	551,971,813	552,131,003
9	275,985,907	276,011,526
10	137,992,953	137,874,262
11	68,996,477	69,127,258
12	34,498,238	34,545,871
13	17,249,119	17,287,151
14	8,624,560	8,645,262
15	4,312,280	4,311,566
16	2,156,140	2,118,205
17	1,078,070	1,055,064
18	539,035	550,366
19	269,517	265,437
20	134,759	132,277
21	67,379	62,715
22	33,690	34,000
23	16,845	18,727
24	8,422	10,795
25	4,211	2,870
26	2,106	1,309
27	1,053	2,118
28	526	521
29	263	0
30	132	0
31	66	0
32	33	0
33	16	0
34	8	0
35	4	0
36	2	0
37	1	0

SYMMETRY AND THE END OF PROBABILITY

The Law of Average as the new Law of Large Number

The Law of Large Number has long been considered as the heart of the Probability theory. Although both the Law of Large Number and the Law of Average predict the convergence of an averaging process, we note from inequality (4) that the Law of Average is superior to the Law of Large Number in two significant ways:

1. The Law of Large Number is a probabilistic law, the Law of Average is not.
2. The Law of Large Number is rather vague, as it can only guarantee that the long-term ratio converges to $p \pm \varepsilon$ where ε is a parameter dependent on N. The Law of Average is much more specific, with $\varepsilon = k\{p(1-p)/N\}^{1/2}$.

However, the main point is this: The Law of Large Number has been mistaken as a result of the Probability theory, but actually is in conflict with it. Since the probability-based proof for the Law of Large Number is invalid, the Law of Large Number as it stands today is no more than an *ad hoc* result devoid of logical value. The Law of Average, on the other hand, is a result of symmetry and is logically consistent.

For this reason we should discard the Law of Large Number as described in textbooks. The Law of Average, then, will take its place as the new and more complete Law of Large Number.

Symmetry and the "Physically Impossible Rationale"

We will now revisit the Physically Impossible Rationale, which states vaguely that events with low enough probabilities are not physically possible, and that is why they do not occur. As we have mentioned in chapters 2, the Physically Impossible Rationale is in conflict with the basic concept of probability, although it has often been used as a quick-fix patch.

From our discussion in the previous section we must conclude that, as it stands today, the Physically Impossible Rationale is not strictly correct. This can be seen by comparing the following combinations in a coin tossing process that prohibits the occurrences of strings with length N+1 and above:

1st combination: N consecutive heads followed by 1 tails

2nd combination: N+1 consecutive heads

The probability of occurrence is 0.5^{N+1} for both combinations. The prohibition threshold must be roughly the probability for N heads, which is 0.5^N. Since 0.5^{N+1} is smaller than 0.5^N, we would conclude that neither the first nor the second combinations could occur; which is a paradox because we know that these two combinations exclude each other, therefore one of them (i.e., the first combination) must occur if enough tosses are attempted.

This paradox is solved by recognizing that, by allowing the occurrence of N consecutive heads or tails but no higher, the process under investigation has a (finite) "randomness capacity" that can only accommodate 2^N different possibilities. When more than N tosses are attempted the randomness capacity is exceeded, and symmetry must lend a helping hand in deciding which combinations are allowed and which combinations are not. The Law of Average achieves this goal by ensuring that all combinations, large and small, occur at their correct frequencies within the allowed asymmetry limit.

Thus, the underlying logic for the Physically Impossible Rationale is again the balance between randomness and symmetry.

The proper way to view the Physically Impossible Rationale, therefore, is to consider it a by-product of the Law of Average, which already includes symmetry. In other words, once we give up the Probability theory, the Physically Impossible Rationale is no longer necessary (though we may still want to keep it for reference purposes.)

Interestingly, the idea of prohibition threshold, which came from the Physically Impossible Rationale, does have an important role in the symmetry approach. We will return to it in a later chapter.

D. FROM PROBABILITY TO DISTRIBUTION

The pseudo-science status of probability

Coin tossing belongs to a class of problem that forces science to deviate from its natural approach. The usual method of describing a process in terms of space and time fails miserably because the outcome of a coin toss does not depend on either of them. A feeling of helplessness must have been the reason why Probability was born, like a child of necessity, to be used and abused but not taken care of. It is puzzling to think that the Probability theory, the foundation of the quantum mechanical breakthrough of the 20th century, is still a pseudo-science today, in the 21st century.

The replacement of Probability theory by Distribution theory

Since the Probability theory is clearly incompatible with long term behavior of distributive processes, we need to replace it with a new theory.

The logic for the new theory is symmetry, which can only manifests its presence in long-term distributions. An appropriate name for the new theory would be "The Symmetry Theory of Distributive

Processes"; but since symmetry can be taken for granted, we will propose a much shorter name "The Distribution Theory".

Distribution theory vs. Probability theory

By utilizing symmetry, the Distribution theory enjoys many significant advantages over the Probability theory.

1. Long-term predictions: The Distribution theory predictions match probability predictions at low sample size and outperform probability predictions at high sample size.

2. Paradoxes and their solutions: The Probability theory is inflicted with many paradoxes, among them the long-run paradox and the Gibbs paradox. The Distribution theory solves these paradoxes with the principle of symmetry. Moreover, the Distribution theory does not create any paradoxes of its own.

3. Agreement with empirically verified laws: The Probability theory violates symmetry and is incompatible with the Law of Large Number. The Distribution theory incorporates symmetry in its foundation and gives rise to the Law of Average, which is a more precise version of the Law of Large Number.

4. Agreement on measure of randomness: The Probability theory assumes infinite randomness, which is in conflict with real life processes such as the process of coin tossing, where at least some outcomes can be analyzed by the deterministic method of Newtonian mechanics. The Distribution theory can handle all mixtures of randomness and determinism.

5. Internal consistency: The "Physically Impossible Rationale" (used by some probability proponents to explain why certain low probability events never take place) is actually in conflict with the foundation of probability. The Distribution theory easily accounts for the absence of "physically impossible" events with the Law of Average. In fact, there is no need for the Physically Impossible Rationale in the framework of the Distribution theory.

The writer believes that the case against the Probability theory is convincing beyond question, and the only logical action is to replace it with the Distribution theory.

The need for new developments for distributions

With the Probability theory out of the picture, we will now focus our attention on distributions. Our concept of distribution is exactly the same as that of the Probability theory. More correctly, we should say both the Probability theory and the Distribution theory apply the same set of statistical principles that govern distributions. We will find, however, that the Probability theory application of statistical distributions is still in a state of underdevelopment with many open questions. The existence of these open questions is a natural consequence of the probability description of distributive processes. Since all unknowns could be considered "probabilistic" there has been no strong motivation for scientists to seek a permanent solutions to problems that show up only intermittently.

We will propose answers to these open questions in the next chapter. Our answers will form the backbone of the Distribution theory.

First written November 2002
Revised December 2002

Chapter 5

Distribution theory I
Finite randomness and inter-dependency of events

> *We will show that the assumption of infinite randomness of the Probability theory is unrealistic, and that is why the Probability theory is incorrect.*
>
> *We will show that certain random combinations can never occur at all. We will call these "prohibited propensities" and provide experimental evidence for them. We will elaborate on the level of randomness factor k (which has been assumed to be infinity and therefore ignored by the Probability theory.) We will prove that the k-factor is constant for ideal systems and approximately constant for efficient artificial systems. We will advance one step ahead of the Probability theory by showing that all distributive processes converge.*
>
> *Most significantly, we will show that individual events in a distributive process cannot be independent of one another as assumed by the Probability theory (and by most of us). We will prove instead that they are inter-dependent. This result is guaranteed to be a shock to the world of science.*

A. THE LOGIC OF FINITE RANDOMNESS

The reality of finite randomness

It is important to differentiate the Probability theory as a branch of mathematics, and the Probability theory as a scientific application. The only requirement of a mathematical theory is that it is self consistent, so the question is not whether the Probability theory is mathematically correct, but whether it is applicable to the real world. This question must be answered by comparing probability predictions with reality.

We have done this in the last few chapters; and we encountered the long-run paradox, the Gibbs paradox, etc. Finally, we were forced to conclude that the Probability theory is not the correct tool to describe distributive processes.

To sum this up, the Probability theory may be a good theory of mathematics[1], but it is definitely not a good theory of science.

Since the applicability problem of the probability theory originates from its Infinite Randomness Assumption, which says that all distributive processes are ruled by infinite randomness, by *reductio ad absurdum* it is a simple matter to conclude that the real world is not ruled by infinite randomness.

In hindsight, the impossibility of infinite randomness is very obvious. Infinite randomness is the same as patternless. It necessarily implies non-convergence and complete unpredictability. Science requires predictability. The fact that we can do science proves that the universe is not ruled by infinite randomness.

New meaning of randomness and the measure of randomness level

Since infinite randomness does not exist, it is necessary to redefine what "randomness" means in the real world.

In the conventional view "randomness" is the same as absolute patternless. This view of randomness has put science and technology in the dark. The ability to produce a long series of heads or tails (0's and 1's in digital language) can be used as a measure of the "randomness level" of a system. By asserting that it is possible for "random processes" to, say, produce an infinitely long series of heads or tails, the Probability theory has guessed wrongly that the randomness level is infinitely large for distributive systems. Scientists were disappointed and alarmed when it was proven that all random number generators would repeat themselves after certain number of trials. This was considered as a proof that human could never imitate the absolute randomness of Nature. We now know Nature is operating just like us, with finite randomness capacity. Finite randomness is not just normal, it has to be the case otherwise science could not exist.

The meaning of randomness, then, will have to be redefined with the added concept of randomness level. Randomness level is a measure of the maximum deviation from the ideal state that a distributive system can achieve. It has to be finite for all distributive processes.

The quasi-continuous nature of distributive processes

We will apply symmetry to a distribution with propensity p, sample size N, and large enough trial size M so that all possible combinations are present in the data set. If we define $R = N_{max} - N_{min}$, then by the results established in the last chapter:

$$R/N = 2k\{p(1-p)/N\}^{1/2} \qquad (1)$$

In classical probability clearly $R/N = 1$. The randomness axiom of frequency probability and the n-freedom approach of Karl Popper[2] also imply that $R/N = 1$.

In earlier chapters we have established that, in the operation of computers, if p (different from 0 and 1) is the long-term propensity of an event, the maximum string of consecutive occurrences of this event must be finite! Let N_{max} be the length of this finite string, since it is impossible to get a string longer than N_{max}, it is obvious that for all $N = N_{max} + n$ where n is a non-zero integer the following is true:

$$R/N < 1 \qquad (2)$$

Inequality (2) contradicts the position held by the Probability theory, but it is fully compatible with equation (1), which is a result of symmetry. Thus, we have confirmed that the Probability theory is not a valid description of computer operation; and we have all reasons to believe that the correct description is symmetry. The same logic is easily extended to distributive systems other than computers.

As seen in the right picture of figure 1, the position R/N=1 taken by classical probability, frequency probability, and the n-freedom approach by Popper all imply that k is infinite. As a consequence, the function R/N is discontinuous at p=0 and p=1.

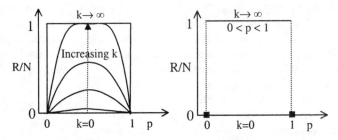

Figure 1: Symmetry (left) leads to quasi-continuous behavior of the R/N function, whereas classical probability, frequency probability, and the propensity interpretation of probability all predict R/N=0 for p=0 or p=1, and R/N=1 for 0<p<1, which is strongly discontinuous (right).

Our approach (symmetry), on the other hand, anticipates that k is finite in the general case, resulting in a family of curves that connects the two points p=0 and p=1 without interruptions (except for the case k = ∞, which corresponds to infinite randomness.) Although each of these curves appears continuous, there exists a small gap between successive points because R and N are finite integers. To be precise we therefore say that R/N is a quasi-continuous function of p.

B. THE k-FACTOR

The randomness level k

The symmetry approach gives rise to the parameter k. This factor is absent in existing Probability theories because it is assumed to be infinity.

The reader may want to convince him or herself that p=0.5 is the only non-trivial solution (p≠0 and p≠1) for k=0. This solution requires heads and tails to alternate each other indefinitely. This is as close to determinism as a distributive process can be, as evident in the sequences shown in table 1.

TABLE 1: Alternating sequence for k=0 (and p=0.5)

Outcome	T*	H	T	H	T	H	T	H	T	H	T
Running p	0.00	0.50	0.33	0.50	0.40	0.50	0.43	0.50	0.44	0.5	0.45

*The first toss is the only unknown in the sequence (it could also be heads)

In the more general case, k reflects the relative strengths of randomness and symmetry. If k is infinity (as assumed by the Probability theory) randomness will overwhelm symmetry, making it completely impossible to guess the next series of n events, including the case n approaching infinity.

From a mathematical standpoint, symmetry can be considered as a fixed force of Nature. It is therefore convenience to think of k as a measure of randomness alone and label it as "the randomness level". Because randomness causes asymmetry, we will also sometimes refer to k by its alternative name "the asymmetry level".

The constant k-factor in ideal processes

We will define an "ideal process" as a process such that when N approaches infinity all combinations considered equivalent by the "equal opportunity principle" are equally represented.

Recall that the equal opportunity principle is a combined effect of randomness and symmetry. Since symmetry is considered a fixed force of Nature, in order for an ideal process to take place the force of randomness has to treat all equivalent combinations equally.

As the name suggests, there is no ideal process in existence. However, the concept of "ideal process" is still helpful because it gives us a reference point to evaluate non-ideal processes.

In binomial processes with large enough N and high enough asymmetry, the distribution curve can be approximated by the standardized normal curve. We note that the range $(-k\sigma, +k\sigma)$ is the range of maximum asymmetry allowed by the process. Since σ is the unit measure of asymmetry, and the maximum asymmetry of the process has to be finite, the process parameter k must be a finite number. In fact, since k corresponds to the maximum (finite) asymmetry, it is by definition a fixed number. In other words, k is a process constant. We will call it "the k-factor" of the process.

According to the Probability theory, the k-factor is infinitely large. Since this does not seem to match reality, some have suggested the idea of "most probable limit". For example, if the most probable limit for one system is 5 sigma's then it is highly unlikely that it will ever wander outside 5 sigma's. The problem is that the Probability theory cannot guarantee that such a system will never exceed 5 sigma's. This kind of workaround is scientifically unsatisfactory. Thanks to the Distribution theory, it is no longer necessary.

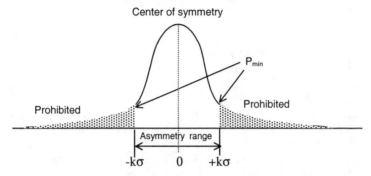

Figure 2: The k-factor is the horizontal distance measured in units of sigma's from the peak of the normal curve to the location of the minimum bias P_{min}. All possibilities are confined in the range +/-kσ. Thus 2kσ is the maximum deviation range. Propensities below P_{min} are prohibited.

(We have exaggerated P_{min} to emphasize the point. It is vanishingly small in most practical cases where the k-factor is sufficiently large.)

Discovering that the k-factor is a process constant is a great breakthrough by the Distribution theory. The logic that we have just presented was deceptively simple, but the route leading to it was a tricky one. The writer will admit that it took him 10 months and approximately 500 hours of thinking time with many false starts, self-contradictory logic, and wrong conclusions.

We are ready to summarize our finding in the next theorem:

THE FINITE LIMIT THEOREM

Given an ideal distribution with propensity p and an averaging process with sample size N large enough for the normal approximation to be valid. Let N_S be the actual outcome of a single trial, then all single-trial deviations from the mean value (i.e., N_S-Np) are confined within $\pm$ kσ; where k is a system constant called the k-factor, σ is the standard deviation, also known as sigma. Mathematically:

$$|N_S\text{-}Np|_{max} = k\sigma = k\{Np(1\text{-}p)\}^{1/2} \qquad (3)$$

The maximum deviation then is identified as kσ.

The k-factor in non-ideal processes

What about non-ideal processes such as the coin tossing simulation performed by a computer? Although a computer is a good imitation of how Nature operates, it is by no means a perfect simulator. For example, when we use the computer to generate random numbers, what we actually get are "pseudo-random numbers", i.e. numbers that appear random but are generated by a known algorithm.

As a rule, all algorithms will (systematically) fail to treat at least a minority of combinations equally. We therefore expect k to be approximately but not perfectly constant for computer simulated processes. Obviously the degree of perfection of artificial systems depends on how well they are constructed (e.g., quality of hardware, software, and firmware).

The bottom line still is, according to the Distribution theory, there exists in every process a (process-dependent but) fixed k-factor that cannot be exceeded, for all propensities p's and sample size N's.

Experimental k-factor results in support of the Distribution theory

For quick reference, we will rewrite equation (1):

$$R/N = 2k\{p(1-p)/N\}^{1/2} \tag{4}$$

Where p is the (long-term) propensity, N the sample size, k the k-factor, and R the difference between the maximum and minimum times that heads occurs in a single trial.

Equation (4) applies only when the trial size M is large enough to warranty that the most extreme combinations occur. Since this condition may not be met in experiments, the valid working condition is:

$$R/N \leq 2k\{p(1-p)/N\}^{1/2} \tag{5}$$

Which can be rewritten in terms of the experimental k-factor denoted as k_E:

$$k_E \equiv R/\{2[Np(1-p)]^{1/2}\} \leq k \tag{6}$$

. According to the Probability theory k is infinity, therefore k_E could be any positive number. Moreover, although the probability for a large value of k_E, say 10 or 20, is extremely low, the Probability theory is forced to hold the position that a large k_E could happen anytime, even in the very next set of trials. The Distribution theory, however, holds the position that k is a finite constant, and k_E cannot exceed k.

Thus, it is possible to show that the Distribution theory is wrong by producing a single occurrence with huge k_E. On the other hand, if many tests with various p's and N's are conducted, and all recorded k_E values are below a reasonable limit k, there are good reasons to believe that the Distribution theory is more precise than, and therefore superior to the Probability theory.

TABLE 2: k_E values for various p, N, M on a system with $k \approx 6$
(test date: September-December 2002)

Samples N	Trials M	P 0.01	P 0.05	P 0.10	P 0.20	P 0.30
2,500	10	1.8	1.5	1.0	2.1	1.6
2,500	100	2.6	2.8	2.6	2.3	2.7
2,500	1,000	3.2	2.9	3.5	3.0	3.0
2,500	10,000	3.7	3.8	3.9	4.0	3.9
5,000	10	2.6	1.3	1.6	1.8	1.3
5,000	100	2.4	3.0	2.6	2.9	2.2
5,000	1,000	3.3	3.4	2.8	3.3	3.2
5,000	10,000	3.7	3.7	3.9	3.7	3.6
10,000	10	0.9	1.5	1.8	1.9	1.5
10,000	100	2.5	2.2	2.6	2.2	2.7
10,000	1,000	3.9	3.0	3.0	3.1	3.0
10,000	10,000	4.1	4.0	4.3	4.1	3.6
100,000	10	1.4	1.4	1.1	1.4	0.8
100,000	100	2.3	2.1	2.2	2.1	2.7
100,000	1,000	3.3	3.0	3.1	3.3	3.2
1,000,000	10	1.0	2.1	2.0	1.3	1.9
1,000,000	100	2.5	2.3	2.2	2.1	2.0

Samples N	Trials M	P 0.40	P 0.50
36	10		1.3
36	100		2.7
36	1,000		3.5
36	10,000		3.7
36	100,000		4.3
36	1,000,000		4.8
36	10,000,000		4.8
2,500	10	2.3	1.3
2,500	100	2.1	2.7
2,500	1,000	3.3	3.3
2,500	10,000	4.0	4.0
5,000	10	1.6	1.3
5,000	100	2.7	2.6
5,000	1,000	3.6	3.2
5,000	10,000	4.0	4.0
10,000	10	1.9	2.0
10,000	100	2.6	2.0
10,000	1,000	3.4	3.2
10,000	10,000	3.7	3.7
100,000	10	1.7	1.2
100,000	100	2.3	3.0
100,000	1,000	3.3	3.7
1,000,000	10	1.7	1.1
1,000,000	100	2.7	2.3

It should be noted that the goal of these k_E tests is to evaluate a prediction made by the Distribution theory but not by the Probability theory. It is therefore not a test for or against the Probability theory. (The best test against the Probability theory is the long-run test, which shows that combinations with very low probabilities are completely absent. This was completed earlier.)

We will discuss a method to determine k from the longest string later. The existing set of data was obtained by a computer simulation process with longest unbiased string of 30, giving k≈6.

Since k=6 corresponds to roughly 2 occurrences in a billion trials, M will have to be at least 10 billion to make sure that the most extreme combinations are exhausted. Due to the time consuming nature of the experiments, this requirement was not met. Therefore, if the Distribution theory is correct, not only none of the reported k_E values can exceed 6, but also most of them should be well below 6.

In table 2 we recorded results of 126 independent experiments (i.e., each data point came from a separate experiment.) All k_E values are well below 6, in agreement with the Distribution theory.

TABLE 3: Average k_E values taken over all sample sizes N's as a function of propensity p and trial size M (for k≈6)
(test date: September-December 2002)

P	0.01	0.05	0.1	0.2	0.3	0.4	0.5
Trial size M							
10	1.5	1.6	1.5	1.7	1.4	1.8	1.4
100	2.5	2.5	2.4	2.3	2.4	2.5	2.5
1,000	3.4	3.1	3.1	3.2	3.1	3.4	3.4
10,000	3.8	3.9	4.0	3.9	3.7	3.9	3.8

TABLE 4: Average k_E values taken over all propensities p's as a function of sample size N and trial size M (for k≈6)
(test date: September-December 2002)

Sample size N	2,500	5,000	10,000	100,000	1,000,000
Trial size M					
10	1.7	1.6	1.6	1.3	1.6
100	2.6	2.6	2.4	2.4	2.3
1,000	3.2	3.3	3.2	3.3	n/a
10,000	3.9	3.8	3.9	n/a	n/a

SYMMETRY AND THE END OF PROBABILITY

In addition, the Distribution theory predicts that the experimental value for k (i.e. k_E) is statistically indifferent to the propensity p and the sample size N (provided that N is large enough so that all physically feasible combinations can be realized). To verify this, we constructed table 3 and 4 from the data of table 2. As the results in table 3 indicate, within tolerance, k_E is indeed indifferent to p. Likewise the results in table 4 indicate that k_E is indifferent to N.

These results confirm the predictions of the Distribution theory; and although they do not refute the Probability theory directly, they are clear indications that the Distribution theory is more precise than the Probability theory. Our call for the end of probability therefore is more than a call for scientific consistency. It is also a practical call for more precision in scientific empiricism.

C. THE INTER-DEPENDENCY OF EVENTS

Process resolution, trial size M, and the convergence axiom of frequency probability

The statistical trend in table 3 indicates that k_E, the experimentally recorded randomness level of the process, is independent of the sample size N but grows with the trial size M. This is not surprising because the more trials are attempted, the more chances there are for randomness to manifest itself.

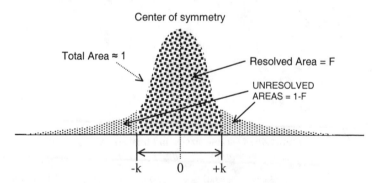

Figure 3: In the standardized normal curve, the horizontal is measured in number of sigma's. It is seen that k sigma's (i.e., the area F under the curve from –k to +k in the picture) fails to "resolve" the areas 1-F at the two tail ends.

But there is a very surprising effect that can be fully appreciated only after we introduce the concept of process resolution.

94

Finite randomness and interdependency of events

The binomial distribution curve constructed with experimental data is like a digitized image formed by the opposing forces of randomness and symmetry. The general shape of this image is decided by the relatively wide area around the peak of the distribution curve, where the effect of deviations is either small or negligible. The finer details are confined to the two narrowed tail ends of the distribution curve, where every deviation is highly noticeable.

We already know that k sigma's include a fraction F of the total area under the distribution curve, and this fraction is centered around the peak of the distribution curve. But this also means k sigma's leave out data at the two tail ends of the distribution curve. Putting it the other way, the fraction of the area under the distribution curve that k sigma's fail to resolve is (1-F).

We will define process resolution as the ratio of the total area of the distribution curve over the unresolved tail-end areas. With this definition, the formula for the process resolution is simply $1/(1-F)$. Assuming that the binomial distribution can be approximated by the normal distribution, we get the following results from standard statistical tables:

TABLE 5: Process resolution vs. k (from statistical tables)

Process Resolution $1/(1-F)$	10	10^2	10^3	10^4	10^5	10^6	10^7
k	1.7	2.6	3.3	3.9	4.4	4.9	5.3

Let's now compare table 5 with table 6, which lists the number of trials M from table 2 versus the average k_E values taken over all sample sizes N. Except for $M=10^7$ -which represents only one data point and therefore is not indicative of the general trend- the striking similarity between k of table 5 and average k_E of table 6 forces the conclusion that M is the process resolution itself!

TABLE 6: Number of trials M vs. average k_E (from data of table 2)
* Cases with single data point (for N=36)

M	10	10^2	10^3	10^4	10^5	10^6	10^7
Average k_E	1.6	2.5	3.2	3.9	4.3*	4.8*	4.8*

In hindsight this only makes sense. The M trials are like the M dots used to digitize the distribution into an image. Since the distribution is fixed by the process, the clarity of the digitized image depends on the dot size. Since M dots are used to digitize the unit area of the standardized distribution curve, the dot size is 1/M. Details smaller than the dot size will not be captured in the digitized image, so 1/M must equal (1-F). It follows that M must equal the process resolution 1/(1-F).

The development of an approximately normal binomial distribution can now be explained. When M is small, we only have a crude outline of the distribution (figure 4, left). When M is relatively large, the bell shape of the distribution curve becomes apparent (figure 4, center). When M is very large, the image is sharply defined (figure 4, right).

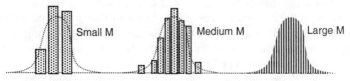

Figure 4: Poorly defined distribution with small M (left).
Reasonably well defined distribution with higher M (center).
Well-defined distribution with very large M (right).

The alert reader will notice that this development mechanism naturally forces all combinations to approach their long-term ratios when M is large enough. Thus, we have just found another logical justification for the convergence axiom of frequency probability[3] by Richard von Mises, and in effect made this axiom unnecessary.

Another symmetry argument against classical probability
Each single trial can also be compared to a qualified dart throw that lands somewhere within the distribution curve, which acts like a target. The only difference so far between this analogy and the Probability theory is that in this analogy k is a finite number, whereas it is infinite in the Probability theory.

But what is the mechanism that makes the long-run number of darts approximately equal everywhere inside the target (i.e., the area under the distribution curve)? More specifically, what prevents the darts from concentrating in one part and missing the rest of the target? The probability answer of course is "equal probability". But, as we have shown over and over many times earlier in this book, this answer would not only lead to contradictions but also imply that an infinitely large number of trials (M→∞) is needed to achieve equal concentration.

Our answer is: Symmetry. Thanks to the fact that the randomness level is finite, symmetry can achieve equal concentration –within certain tolerance- rather quickly. This fits experimental observations.

Symmetry is a comparative mechanism. If there exists asymmetry in the system, symmetry must react accordingly so that the level of asymmetry does not grow out of control. But in order to do this, it cannot treat all future trials equally. Events that should have occurred more often will have higher probability to occur in the future, while events that should have happened less often will have lower probability (see figure 5).

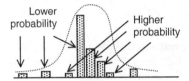

Figure 5: The Principle of Symmetry implies that probabilities of future events may depend on what has happened in the past. This violates the Indifference Principle of probability

Return to our dart-and-target analogy. Symmetry is like a house rule that doesn't allow the dart thrower to hit the same spot, say, three times in a thousand trials. If he has hit one spot once, he is still allowed to hit it once more, but if he has hit the same spot twice, the spot will be blocked off so that he cannot hit it again. After a thousand trials, there will be spots with no hit, with one hit, with two hits, but no spot with three or more hits. It is only with this kind of mechanism that a reasonable long-term balance between randomness and symmetry can be guaranteed.

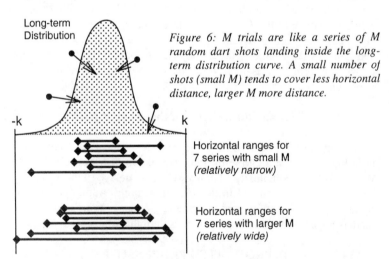

Figure 6: M trials are like a series of M random dart shots landing inside the long-term distribution curve. A small number of shots (small M) tends to cover less horizontal distance, larger M more distance.

Horizontal ranges for 7 series with small M *(relatively narrow)*

Horizontal ranges for 7 series with larger M *(relatively wide)*

Before the first dart is thrown, the target is empty and all combinations should have equal probability; but once a number of trials have been attempted, they will affect the probability values of future throws! Needless to say, this is a blatant violation of the indifference principle, on which the fixed probability assumption of classical probability is built on!!! Essentially symmetry has forced us to discard the (classical probability) assumption that each single event (in a series of events) has the same probability.

Thus, if the concept of probability is to be applied, probability must vary from event to event. Specifically, there must exist cases where the probability for an event to happen is zero (i.e., it cannot happen at all), and other cases where it is 1 (i.e., it is certain to happen).

The interdependency of individual events in distributive processes

The shocking implication is that individual events in distributions cannot be considered as separate entities, but have to be connected into groups of multiple events. In other words, individual events in a distributive process are not independent of one another as alleged by the "indifference principle" of probability. Instead, the reality of finite randomness requires them to be inter-dependent!!!!

By proving that individual events are interdependent in distributive processes, we have removed the last reason to appeal to (fixed) event probability in the analysis of distributions. Thus, for all practical purposes, the idea of fixed probability is a relic of the past in the Distribution theory[3].

The interdependency of single events gives us a completely new and revolutionary view of all branches of knowledge dealing with groups of elements, not only the (innate) natural sciences but also the so-called "fields of life" such as psychology, sociology, anthropology, evolution, and genetics. Regrettably, these subjects are outside the scope of this book. The writer hopes to have a chance to discuss them in a future opportunity.

Inter-dependency and the meaning of time in distributive processes

Since individual events in most distributive processes follow one another in the dimension of time, the inter-dependency of events brings up a big issue regarding the meaning of time in distributive processes. Because of the complexity of this issue, it is delayed until chapter 7, where it will be discussed in the special section: "Additional reading: Symmetry, synchronicity, and the meaning of space-time". The curious readers are asked to hold their questions until that time.

D. PROHIBITED PROPENSITIES

The symmetry logic of prohibition threshold

In chapter 2 we introduced the concept of the prohibition threshold P_{min} as a consequence of the Physically Impossible Rationale, then showed that this concept is incompatible with the fundamental principle of probability. The reason is that the probability theory must allow all possibilities to take place, but according to the concept of prohibition threshold, events with probabilities below P_{min} cannot take place.

There is no such problem in the Distribution theory because prohibition threshold is a natural consequence of symmetry. It is simply the maximum asymmetry allowed by the distributive process under investigation. In a coin tossing process, prohibition threshold corresponds to the maximum consecutive occurrences of heads (or tails) allowed by the process. We will see practical examples of prohibition threshold later.

Since the area under the standardized normal curve is exactly 1, the location for the prohibition threshold P_{min} is at the location where it is equal to the vertical value of the distribution curve. Since the ideal normal curve is symmetrical with respect to its peak, the location of P_{min} corresponds to $k\sigma$'s from the peak of the curve (which serves as the center of symmetry of the process.) Since k is fixed for ideal processes, P_{min} will also has to be fixed.

In real processes, which are non-ideal, we expect P_{min} to be approximately constant.

Prohibited propensities

On the average, the count for an event with propensity p is X when the sample size is $N=X/p$. In reality after X/p trials the event will occur somewhere between R_1 times and R_2 times, where $R_1 < X < R_1$. The range (R_1, R_2) is the maximum deviation of the propensity p.

Obviously, the range (R_1, R_2) increases when the sample size N increases. Since the theory of Probability assumes infinite randomness, it demands that $(R_1, R_2)=N$. In the more realistic case of finite randomness, both (R_1, R_2) and N have to be finite. This means N cannot exceed a maximum system limit N_0. The minimum propensity is the propensity that can occur once with sample size N_0:

$$p = 1/N_0 \qquad (4)$$

The reader may have noticed that "minimum propensity" is just a more descriptive name for "prohibition threshold". Since the value for p is prohibited from going lower than $1/N_0$, we define the minimum propensity:

$$P_{min} \equiv 1/N_0 \qquad (5)$$

Mathematically, propensities lower than P_{min} can never occur. They are the prohibited propensities

$$p \text{ (prohibited)} < 1/N_0 \qquad (6)$$

"Prohibited propensities" is a crucial departure from the Probability theory, which claims that all possibilities will be presented in the final distribution. For example, if we flip one million unbiased coins, the inherent propensity for all coins to turn up heads is $1/2^{1000000}$, an extremely small but non zero number. According to the Probability theory, this extreme combination will occur if enough trials are attempted; but according to the Distribution theory, it is prohibited in realistic processes, and therefore we will never see it, even in the theoretical case of infinite number of trials.

Prohibited propensity and system capacity

A qualification is necessary before we go on. Earlier we said that the sample size has to be finite in a distributive process. This does not mean a process has to stop after a finite number of events. It simply means if the sample size goes beyond a maximum limit N_0, the system will start repeating itself. To avoid this from happening, the capacity of the system has to be many times greater than N_0. We define the capacity of the system as:

$$U = KN_0 = K/P_{min} \qquad (7)$$

Where K is a system constant. Solving for P_{min}:

$$P_{min} = K/U \qquad (8)$$

Thus, in practice only a part of system capacity is used to accommodate P_{min}. This is the rule in all random number generators. For example, the capacity of the random number generator used in most simulation tests for this book is 10^{15} (i.e., producing random number with 15 significant digits); but the minimum propensity is believed to be in order of 10^{-8}, which is larger than $1/U$ by about 10^7 times.

More experimental verification of prohibited propensity

In chapters 1 through 3, we presented the same set of data for a coin tossing simulation experiment performed on a Macintosh computer with propensity p=0.50, where strings of length 29 or more consecutive heads or tails never occur. These are real life examples of prohibited propensities.

We will now present additional data taken on a window-based computer. This set of data cover p values of 0.05, 0.10, 0.2, 0.3, 0.4, and 0.5.

Due to the lack of mathematical symbols, we will use the computer symbol "^" for the power operation, i.e., x^y means "x to the power of y".

It follows from (8) that P_{min} is a system constant; meaning that it may vary from system to system, but is fixed for a given system. To verify (8) experimentally we note that the propensity for consecutive occurrences of N identical single events is $P_{CP} = p$^N, where p is the propensity of each single event. There exists a sample size N_{max} so that

$(P_{CP})= p^{\wedge} N_{max}$ is barely larger than the prohibition threshold. We will use the combined symbols ">≈" to mean "barely greater".

$$p^{\wedge} N_{max} >\approx P_{min} \tag{9}$$

If we choose the largest p as the reference propensity p_0, then its corresponding maximum consecutive occurrences $(N_0)_{max}$ will also be the largest of all N_{max}'s. An arbitrary single events "1" is then related to the reference event "0" by the relationship:

$$p_1^{\wedge}(N_1)_{max} \approx p_0^{\wedge}(N_0)_{max} \tag{10}$$

Since all p's are smaller than 1, (10) leads to:

$$(N_1)_{max} \approx (N_0)_{max}\{\ln[p_0]/\ln[p_1]\} \tag{11}$$

Where "ln(x)" means the natural logarithm of x. Note that all N_{max}'s have to be integers.

Relationship (11) can be verified against experimental data by the following procedure:

1. Set p and search for N_{max} (by running long series of numbers repeatedly and look for maximum consecutive occurrences.) One simple way to simulate this test on a computer is to generate a long series of random numbers with a chosen propensity p and convert them to 0's and 1's, then search for consecutive strings of the same.
2. Change p and repeat 1.
3. Choose the largest N_{max} as $(N_0)_{max}$ to compare experimental data against (11).

The reason for the choice of the largest N_{max} as the reference $(N_0)_{max}$ is to reduce calculation round-off error. Test results are listed in table 7.

TABLE 7: Maximum consecutive occurrences N_{max}
(Test date: August 2001 for p=0.1 to 0.5. November 2001 for p=0.05)

Individual propensity p	Combined propensity $p^{\wedge}N_{max}$	Calculated N_{max}	Actual N_{max}	Error
0.05	0.000000313	5	5	0
0.10	0.000000100	7	7	0
0.20	0.000000102	10	10	0
0.30	0.000000048	14	12	-2
0.40	0.000000069	18	16	-2
*0.50	0.000000030	*25	*25	*0

* reference propensity p_0 =0.5, maximum consecutive occurrences 25

The difficulty of this test is in step 1. Without carrying out the time consuming process of performing each experiment with M of at least 10 billion, it is almost impossible to know with absolute certainty whether the longest string observed is actually N_{max}. In addition, it is unclear if the random number generator treats all combinations equally (this is why the name of the generator was not mentioned.) Nevertheless, the agreement was reasonably good. Readers who are fluent with computer programming may want to repeat this test to verify to him or herself that prohibited propensities do exist in distributive processes.

Some implications of prohibited propensities

One very interesting case of prohibited propensity is when $p < P_{min}$ and $N=1$. The shocking result is that although p is non-zero, it still cannot occur! A real life example is a first generation robot arm designed to achieve heads with every toss. In this case, the prohibited propensity belongs to tails. Each toss would wobble a little, but the robot arm somehow forces the coin to turn up heads consistently. *(If the wobbling worsens, however, at some point p will be marginally equal to P_{min}, and tails will show up intermittently. This situation is usually referred to as "threshold to turbulence", "threshold to randomness", or in more modern language "threshold to chaos".)*

In processes where randomness is not wanted at all, "prohibited propensities" manifest themselves through harmless system noises. The shaking of the daily train at an intersection of two railroad tracks, for example, may give a new passenger a scare, but is no more than a nuisance for the seasoned travelers because they know the "overturn" possibility is prohibited from happening.

Extreme k-factor and determinism

With maximum string of 25 for individual propensity p=0.50 (last row in table 7), the prohibition threshold P_{min} is estimated at $0.5^{25}=3\times10^{-8}$. Using tables for the standardized normal curve, the corresponding k-factor is 5.73. The following table lists maximum consecutive occurrences as function of p and the k-factor. The reader can verify that the calculations match those in table 7 for the case k=5.73.

Some readers may be curious to know if it is actually possible to get one million or one billion heads in a row like the Probability theory has claimed. Surprisingly, this is theoretically possible; and there are even more than one way to do it.

The first way is close to what the Probability theory had in mind: Increasing the k-factor to approach total randomness. An unbiased coin could potentially turn up one million heads or tails in a row when the k factor is 1177.41 (shown in **bold** in table 9). To appreciate how ridiculous this number is, the k-factors for most real life processes are between 3 and 6. It is therefore safe to say that we will not see any man-made system produce one million consecutive unbiased 0's or 1's for a long time to come.

TABLE 8: (Calculated) Maximum consecutive occurrences N_{max}

k-factor	3	4	5	**5.73**	7	8
p						
0.0001	0	0	1	1	2	3
0.001	0	1	1	2	3	4
0.01	1	1	2	3	5	7
0.05	1	2	4	**5**	8	10
0.10	2	3	5	**7**	11	14
0.20	3	5	8	**10**	15	20
0.30	4	7	11	**14**	21	27
0.40	5	9	14	**18**	27	35
0.50	7	12	19	**25**	36	47
0.60	10	17	26	33	49	64
0.70	15	25	37	48	71	92
0.80	24	39	60	77	113	147
0.90	51	84	127	164	241	312
0.95	105	173	261	337	495	641
0.99	539	887	1,335	1,724	2,529	3,275

TABLE 9: Extreme maximum consecutive occurrences N_{max}

k-factor	**2**	250	500	750.00	**1177.41**
propensity					
0.0001	0	3,393	13,572	30,536	75,258
0.001	0	4,524	18,096	40,715	100,344
0.01	1	6,786	27,144	61,073	150,515
0.05	1	10,432	41,726	93,884	231,379
0.10	1	13,572	54,287	122,146	301,031
0.20	2	19,417	77,667	174,751	430,678
0.30	2	25,957	103,824	233,602	575,718
0.40	3	34,106	136,421	306,945	756,473
0.50	4	45,086	180,338	405,759	**1,000,003**
0.60	6	61,177	244,704	550,581	1,356,919
0.70	8	87,618	350,462	788,536	1,943,363
0.80	13	140,049	560,182	1,260,404	3,106,292
0.90	28	296,610	1,186,412	2,669,416	6,578,831
0.95	58	609,261	2,436,985	5,483,192	13,513,443
0.99	298	3,109,448	12,437,495	27,984,239	68,967,745
0.999997	**999,999**				

SYMMETRY AND THE END OF PROBABILITY

The second way is very achievable. In fact it works with any non-zero k-factor value. The writer has purposely chosen k=2 because this value is lower than most practical processes. One million heads in a row is possible when the propensity p is larger than 0.999997, which is extremely close to 1.

But what does it means with p being so close to 1? Determinism, of course! Table 3 shows clearly that, at least numerically, extreme randomness is not the reverse of determinism. In fact, by looking at the bottom right corner of both tables 8 and 9, we can confirm that the two reinforce each other. This shocking equivalence between determinism (high p) and randomness (high k-factor) brings up a mind-boggling, even philosophical, question: Could the apparent determinism that we see in the universe (and upon which classical physics was based on) is simply a part of a long string in an extremely random process? Current science cannot answer this question. The writer hopes that he has the chance to return to it in a future opportunity.

E. CONVERGENCE CONDITIONS

Convergence limit and convergence theorem

For convenience we will normalize sigma by dividing it by the mean value:

$$\sigma_M = \sigma/\mu = \{Np(1\text{-}p)\}^{1/2}/(Np) = \{(1\text{-}p)/(Np)\}^{1/2} \qquad (12)$$

Since the normalized sigma is used more frequently than the original sigma σ, from now on we will refer to it simply as "sigma". All deviations will be measured in units of sigma.

The maximum deviation, $|N_S\text{-}Np|_{max}$ where N_S is the outcome of a single trial, can be normalized the same way by defining:

$$\varepsilon_P \equiv |N_S\text{-}Np|_{max}/\mu. \qquad (13)$$

The equations for the normalized deviation then is:

$$\varepsilon_P = k\sigma_M = k\{(1\text{-}p)/(Np)\}^{1/2} \qquad (14)$$

Note that all deviations are confined within k sigma's. Alternatively, we say all p's converge to within ε_P. For this reason we will call ε_P the "convergence limit" of p.

Just from the form of (14) we see right away that convergence is not possible at all with infinite randomness, because the k-factor would be infinity. This was one major problem that the probability theory could not resolve because it already fell into the trap of infinite randomness.

Equation (14) marks two achievements by the Distribution theory:
1. It is the mathematical statement of a convergence theorem.
2. It confirms once more that the assumption of infinite randomness is the main reason why it is impossible to guarantee convergence in the Probability theory.

CONVERGENCE THEOREM:

All allowed combinations of inherent propensity p in a distributive process of sample size N will converge with respect to N according to the equation:.

$\varepsilon_P = k\{(1-p)/(Np)\}^{1/2}$

Where ε_P is the convergence limit, and k a system constant called the k-factor.

The reality of convergence in real processes is a verification of the Distribution theory's position of finite randomness, and a negation of the Probability theory's position of infinite randomness. We can now state with confidence that the Distribution theory is the correct description of distributive processes, and the Probability theory is not.

Convergence and convergence failure

"Convergence" is actually a subjective concept because we could arbitrarily choose a "convergence limit" ε_{P1} and say that the propensity p converges if it is possible to make $\varepsilon_P < \varepsilon_{P1}$, fails to converge otherwise.

Obviously ε_{P1} corresponds to the minimum acceptable propensity p_1 and requires the maximum possible number of trials U to achieve. Note that U is the system capacity that we have mentioned earlier. Note also that, being a subjective choice, p_1 does not have to be the same as the prohibition threshold P_{min}, but it has to be at least P_{min} in order to occur in the process.

p_1 must be a very small number compared to 1, so $(1-p_1)^{1/2}$ must be almost 1. Therefore:

$\varepsilon_{P1} = k\{(1-p_1)/(Up_1)\}^{1/2} \approx k/(Up_1)^{1/2}$ (15)

Where the symbol "$\approx$" stands for "almost equal to". Divide (14) by (15), we get:

$\varepsilon_P/\varepsilon_{P1} \approx \{(1-p)(Up_1)/(Np)\}^{1/2}$ (16)

It is clear from (16) that if $p>p_1$ we can always achieve convergence (i.e., making $\varepsilon_P < \varepsilon_{P1}$) by increasing the number of trials N. That is because, as long as $N< U$, the following always holds true:

$N \geq U(1-p)(p_1/p)$ (17)

The more interesting case is $p < p_1$. Since no further improvement in convergence can be achieved by increasing the number of trials beyond U, the maximum number of meaningful trials is U. Replacing N by U and noting that now $(1-p)^{1/2} \approx 1$:

$\varepsilon_P \approx \varepsilon_{P1}(p_1/p)^{1/2}$ (18)

Since p_1/p is greater than 1, it is always true that $\varepsilon_P > \varepsilon_{P1}$; thus all combinations with $p < p_1$ will fail to converge to ε_{P1}.

Convergence size

Recall the convergence condition:

$$\varepsilon_P = k\{(1-p)/(Np)\}^{1/2} \tag{19}$$

Solving for N:

$$N = (k/\varepsilon_P)^2\{(1-p)/p\} \tag{20}$$

We define the convergence size N_C to be the sample size N required to reach a desired convergence limit ε.

$$N_C = (k/\varepsilon_P)^2\{(1-p)/p\} \tag{21}$$

Take the case of a coin tossing process with k=3 (three sigma's) and p =0.5. If $\varepsilon_P = 0.01$ (converging to within 1%). By equation (21) the convergence size is:

$$N_C = (3/0.01)^2(1-0.5)/0.5 = 90,000$$

Thus, for this particular process, converge to within 1% is achieved with sample size equal to or greater than 90,000.

The next chapter

This chapter was very much dedicated to the nature of randomness. We showed that randomness is necessarily finite, otherwise there would be no convergence. On the other hand, convergence is guaranteed for propensities that are allowed to occur, as long as they are larger than a chosen limit. In the next chapter, we will investigate the most important theorem of distributive processes, namely the Central Limit Theorem.

First written January 2001
Revised August 2001, November 2001, December 2002
©DangSon. All rights reserved

NOTES

1. Even the mathematical theory of probability needs fixing, because the Law of Large Number must be modified with symmetry. The question is why bother keeping the concept of probability, why not just use symmetry as the founding principle and choose a new name to indicate this change.

2. The concept of n-freedom is discussed at length by Karl Popper in his classic book: "The logic of scientific discovery", Karl Popper, English translation 1965, Harper Torchbooks. This is a revised version of the original in German (published 1935) and earlier English editions (1959, 1960).

3. The convergence axiom advocated by Richard von Mises is basically this: "The long-term frequency of an event approaches its expected long-term ratio when the sequence is long enough". This axiom is an *ad hoc* statement with no logical justification. In fact, even the meaning of "long enough sequence" is ill-defined.

4. The possible future theory of variable probability

The reality of variable event probability contradicts classical and frequency probability as well as the Popperian propensity interpretation of probability as these theories either assume or imply fixed probabilities. Interestingly, subjective probability is not affected by, and in fact is in agreement with, this reality.

Although the name "subjective probability" is an accurate description, it has bad connotation as being "subjective" may be considered to be the same as being "non-scientific" or even "un-scientific". With anticipation that this interpretation of probability will eventually become an important scientific subject, the writer will take the liberty to change its name to "the variable Probability theory", which may be abbreviated as "variable probability" for convenience.

The scientific development of the variable Probability theory is a new challenge that will require a lot of insights and efforts. The writer hopes to present his works on this subject in a near future.

5. Performance of random number generators

Question: What does it means if a random number generator can never produce strings higher than, say, ten unbiased 0's or 1's? Is it worse or better than the one that can produce strings of, say, thirty unbiased 0's or 1's.

Answer: The level of randomness of the first generator is relatively low, but it may still qualify as a "good" random number generator! It will not be the appropriate tool to simulate, say, 25 coins at a time; but it would do fine and may even be the most fitting tool for simulations experiments of, say, 5 coins at a time. The second generator should work well with simulation experiments for up to 30 coins, but it may converge too slowly for experiments with, say, 5 coins. Each generator has its own domain of optimized application. We have to choose the appropriate generator for our purpose.

In other words, randomness level should be evaluated as fitting or not fitting a given purpose. A "more random" generator is not necessarily "better" than a "less random" generator.

Chapter 6

The foundation of Distribution theory II

The binomial nature of the Central Limit Theorem

We will show that the Law of Average and the Central Limit Theorem averaging process are identically the same. We will reach the conclusion that the Central Limit Theorem is the governing rule of all distributive processes.

We will provide a new and much simpler proof for the Central Limit Theorem to bring out its binomial nature. We will also make one correction to the existing proof of the Central Limit Theorem (that requires the sample size to approach infinity, which causes absurdity) and give the Central Limit Theorem more practical value by providing criteria for sample size and trial size for averaging processes.

As a side result, we will provide experimental evidence to show that the sigma formula for binomial processes works even with very small sample size.

A. THE MEANING OF VARIANCE
AND SIGMA IN ARBITRARY DISTRIBUTIONS

Sigma is an important parameter in analyses of actual data. To avoid confusion, we will follow the convention that we established in an earlier chapter and call each data point a trial, not a sample. Recall that the symbol for samples is N, for trials is M.

The calculation of sigma starts with the mean value μ, which is the average value of all data points.

$$\mu = (\Sigma x_i)/M \tag{1}$$

Next is the standard variance, which is the average of the squares of all deviations from the mean value μ:

$$S = \Sigma \Delta_i^2/(M-1) = \Sigma(x_i-\mu)^2/(M-1) \tag{2}$$

Note that the denominator is (M-1), not M. This comes from a practical consideration. Since a set of actual data has no reference point to start with, one of the data point has to play this role, and therefore cannot participate in the averaging process.

Sigma is defined as the square root of the variance:

$$\sigma \equiv S^{1/2} \tag{3}$$

Definition (3) applies to all distributions, so it must apply to binomial distributions. We will verify that this is indeed the case.

The binomial nature of the Central Limit Theorem

The main idea is that the sigma for a binomial distribution does not change with each trial because both N and p should be fixed. However, each trial only gives us one value for the deviation. Since sigma is the average deviation, we have to establish it by averaging all individual deviations.

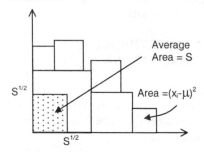

Figure 1: Standard deviation calculations treat a process as if it is binomial. The trick is to move from the linear space to the "areal space" then come back to the linear space. The procedure is squaring the deviations, averaging them, then taking the square root.

But what is the correct averaging method for individual deviations? We already know that we cannot use their numerical averages. However, we do know that each deviation must be of the form $\{N_ip(1-p)\}^{1/2}$, which is the binomial average of $N_ip(1-p)$. We also know that the correct form for sigma is $\{Np(1-p)\}^{1/2}$. Our goal is therefore to first reconstruct $Np(1-p)$ from the individual $N_ip(1-p)$'s by an averaging process appropriate for them.

The appropriate averaging process for $N_ip(1-p)$ is numerical average, because all N_i's are in the same environment. The product $p(1-p)$, on the other hand, is a constant C because p is by definition the long-term propensity.

The formula for N then is:

$$CN = Np(1-p) = C \, \Sigma N_i /(M-1) \qquad (4)$$

Where the summation operation (Σ) is taken over all M trials. We have used (M-1) instead of M in the denominator for reasons cited earlier.

On the other hand, the absolute value of a data point "i" is:

$$|\Delta_i| = \{N_ip(1-p)\}^{1/2} = C^{1/2}N_i^{1/2} \qquad (5)$$

Equations (4) and (5) give:

$$Np(1-p) = \Sigma \Delta_i^2/(M-1) \qquad (6)$$

By comparing (2) and (6) we find:

$$Np(1-p) = S \qquad (7)$$

This leads to:

$$\sigma = \{Np(1-p)\}^{1/2} = S^{1/2} \qquad (8)$$

Which is exactly the same as the result for the binomial sigma that we derived in chapter 4.

We already know that binomial distributions with large enough N are approximately normal, and a normal distribution is fully characterized by mean and sigma. We will see later that many real life distributions happen to be approximately normal. This makes equations (1) through (3) extremely important in data analysis as they allow us to calculate mean and sigma directly from raw data.

B. THE CENTRAL LIMIT THEOREM
and ITS BINOMIAL NATURE

The Central Limit Theorem (CLT)

The Central Limit theorem (abbreviated CLT) is one of the most thought-provoking theorems of all branches of mathematics. In simple terms, the statement of the theorem is:

Let A be an arbitrary distribution with finite mean value μ and finite standard deviation σ. Let B be the "average" distribution made up by samples of size N taken randomly from A. When N approaches infinity B approaches a normal distribution with mean value μ and standard deviation σ $/N^{1/2}$.

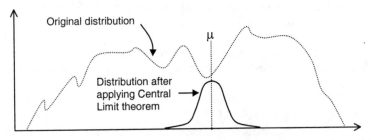

Figure 2: The Central Limit theorem is definitely one of the most beautiful theorem in mathematics. It transforms (almost) everything to a normal curve. Here a very random looking distribution has been transformed to a Gaussian curve thanks to the action of the Central Limit Theorem. Surprisingly there are not that many applications for this powerful theorem. But that will change.

Imagine a pile of dust particles with an irregular distribution of particle size. Simply by taking samples of 100 particles each from this pile and plotting their average size, we find all irregularities have disappeared; and we get a beautifully symmetric bell-shaped curve of the normal distribution! The Central Limit theorem works like a magical wand, putting disorder into order. No wonder it is one of the most thought provoking theorems of mathematics.

We will soon see that the Central Limit theorem is the governing rule of all distributive processes.

The binomial nature of many samples
(new proof for the Central Limit Theorem)

Let's focus our attention on an arbitrary distribution, which we call A. Since this distribution is arbitrary, it may not be binomial. Let x_{min} and x_{max} the smallest and largest values of all samples. We will arbitrarily set:

$$x_{min} = 0 \tag{9}$$
$$x_{max} = N \tag{10}$$

The average value, then, will be somewhere between 0 and N. We will call it Np, where p is a fictitious binomial propensity that we will have use for later. Obviously p is less than 1.

Our actions are unjustified only if x_{min} is minus infinity, or x_{max} is plus infinity. However, since we have asserted in an earlier chapter that infinities do not exist in real life, we will not be concerned with these possibilities.

Now we will perform M random trials. In reach trial, we take N samples from distribution A, average their values, and use the results as new data points to build a new distribution B. It can be seen that the possible range for B is still 0 to N.

We will state without proof that after sufficient trials, the new distribution B will have the same average value as the original distribution A, namely Np. The reader may want to prove this as an exercise.

Mathematically our averaging action (to form distribution B) can be expressed as:

$$X = \Sigma x_i / N \tag{11}$$

Where X is the average value of the N individual samples x_i's.

We notice from (11) that distribution B is the exact distribution that led to the Law of Large Number (appendix, chapter 2). From chapter 4, we know that the Law of Large Number is the same as the Law of Average, which is governed by symmetry and gives rise to a binomial distribution with sample size N. It follows that B is approximately normal when N is large enough. We will quantify "large enough" later.

What about the value of sigma for distribution B? Recall that standard deviation is the sigma value of a binomial distribution which is equivalent to the arbitrary distribution under investigation. So although distribution A is arbitrary, its standard deviation still has the form:

$$\sigma_A = \{N_A p(1-p)\}^{1/2} \tag{12}$$

Where N_A is the (fictitious) sample size of A. Since the sample size of distribution B is N times larger than A and the same propensity p, had it not been averaged its sigma would be:

$$\sigma'_B = \{N_B p(1-p)\}^{1/2} = \{NN_A p(1-p)\}^{1/2} \tag{13}$$

However, due to the averaging process, the true sigma for B is reduced by 1/N, giving:

$$\sigma_B = \{NN_A p(1-p)\}^{1/2}/N = \{N_A p(1-p)\}^{1/2}/N^{1/2} \tag{14}$$

Which is the same as:

$$\sigma_B = \sigma_A /N^{1/2} \tag{15}$$

The alert reader must have noticed that we have just completed a deceptively simple proof for the Central Limit Theorem, with a new emphasis on binomial process.

The method of Fourier transforms is the core for all textbook proofs of the Central Limit theorem. The mathematics of these proofs is elegant but the physical meaning of the theorem is lost in them. The beauty of the Central Limit theorem lies in its ability to bring out the hidden binomial nature of the universe. This important point was lost in the complex mathematics employed by the probability theory. We finally were able to bring out this point in the new proof.

An adjustment to the Central Limit theorem is necessary to make it consistent with the Distribution theory. The textbook Fourier transform proof for the Central Limit Theorem works only if N is infinitely large. Since σ_A is fixed, an infinitely large N would imply that σ_B approaches zero. This is impossible because if σ_B is zero, p has to be either 0 or 1 (total determinism). In reality, p has already been fixed by distribution A.

This theoretical problem is removed by realizing that the Central Limit Theorem averaging process has to be performed by real systems, and all real systems have finite randomness capacity. Thus, N cannot be infinitely large. As a practical result, the resulting distribution is not exactly normal. It is approximately normal with a finite number k of sigma's, which reflects the level of randomness of the system in operation.

As an intellectual exercise, the reader may want to verify to him or herself that the Central Limit Theorem averaging process will fail to converge if the randomness level is infinitely large. This is another important point that the probability theory has missed completely.

The Central Limit Theorem
as the convergence theorem for distributive processes

We must keep reminding ourselves that if single event probability exists, it cannot be fixed for every single event. We use the concept of inherent propensity, which is derived from the final distribution, only as a tool to help us have a feeling of "here and now" and individuality when we deal with a single event. Thus, when we talk of the inherent propensity of a coin, we mean an average propensity of many coins. We should not confuse inherent propensity as something inherent with every coin. In fact, the same coin may even have different propensities in two different processes.

Let's now return to our familiar example of a series of coins being tossed, then turning up as a combination of heads and tails. If we perform M trials, each time with N samples; then keep increasing N and repeat more trials we will discover that they do not vary wildly, but seem to be restricted within a limit, and that limit has something to do with N.

We already know that this limit is the result of the equal opportunity principle, which is a compromise between randomness and the Law of Average (which is the mathematical manifestation of symmetry.) Let's now examine how the Law of Average works.

Mathematically each coin is a trial of sample size 1. Since the inherent propensity is fixed at p, the binomial sigma for each coin is:

$$\sigma_1 = \{N'p(1-p)\}^{1/2} = \{(1)p(1-p)\}^{1/2} \tag{16}$$

Let's examine the situation after N coins. If we take the average, then by the Central Limit theorem the overall sigma has been reduced to:

$$\sigma'_N = \{p(1-p)\}^{1/2} / N^{1/2} \tag{17}$$

But since we are dealing with all N coins, we must multiply σ'_N by N to get the true sigma:

$$\sigma_N = N\{p(1-p)\}^{1/2} / N^{1/2} \tag{18}$$

This gives us the result:

$$\sigma_N = \{Np(1-p)\}^{1/2} \tag{19}$$

Which is the familiar formula for sigma of sample size N and inherent propensity p.

We have learned earlier in chapter 4 that the same formula for sigma is the result of the Law of Average. We have just obtained the same result with the Central Limit Theorem. Thus we have confirmed once more that the Central Limit Theorem is no other than the working mechanism of the Law of Average.

Since the Law of Average is a part of the equal opportunity principle, we have thus confirmed that the equal opportunity principle naturally leads to the Central Limit Theorem. We also confirmed that the convergence mechanism of all distributive processes is governed by the Central Limit Theorem, and by the Central Limit Theorem alone. In other words, the world of distributive processes is ruled by the Central Limit Theorem.

The Probability theory treats the Central Limit Theorem very lightly. This is understandable because it <u>is</u> a convergence theorem, and no convergence could be assumed under the condition of infinite randomness. Now that we have realized that infinite randomness is incompatible with reality, and the mechanism of convergence is guaranteed thanks to the averaging action described by the Central Limit Theorem, we have finally brought respectability and justice to this great theorem.

Sample size requirement for the Central Limit Theorem

If we are faced with an arbitrary distribution and want to use the averaging process of the Central Limit Theorem to create an approximately distribution out of it, what is the sample size N that we should use? This is a very practical question. The condition for the Fourier transform proof of the Central Limit Theorem states that N has to be "arbitrarily large", approaching infinity. In practice, however, experimentalists found in most cases N between 10 and 30 would work very well. In addition, they found a skewed distribution (i.e., p and 1-p are significantly different) would require bigger N than a balanced distribution (i.e., p and 1-p are approximately the same.) Let's explain why this is the case.

Just to be complete, let's start with the trivial case where the original distribution is already approximately normal. The reader may want to convince him or herself that any sample size, including N=1, will work for this special case. The following analysis is focused on the more general case of non-normal starting distributions.

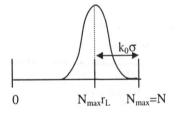

Figure 3: The sample size N_{CLT} required for a CLT process has to obey only one rule: It must allow an approximately normal distribution to develop fully.

We have learn earlier that there is no theoretical problem in creating a CLT distribution with sample size N and mean value Np. Also, there is no theoretical problem in forcing p to be larger than (1-p). For example, if p<1-p in the original distribution A, we can mathematically create a new distribution A' by defining:

$$(x_{A'})_i \equiv (x_A)_{max} - (x_A)_i$$

The binomial nature of the Central Limit Theorem

Where $(x_{A'})_i$ and $(x_A)_i$ are the value of samples "i" in A' and A respectively, and $(x_A)_{max}$ the maximum value of all samples in A. This simple operation switches p and (1-p).

The point is that we do not have to worry whether p is larger than (1-p), we only need to identify r_L and r_S as the larger and smaller of them.

Recall from the last chapter that maximum deviation is related to sigma and the k-factor by the equation:

$$|N_S-Np|_{max} = k\sigma = k\{Np(1-p)\}^{1/2} \qquad (20)$$

where N_S is the single-trial outcome. We will replace p by r_L to get:

$$|N_S-Nr_L|_{max} = k\{Nr_L(1- r_L)\}^{1/2} \qquad (21)$$

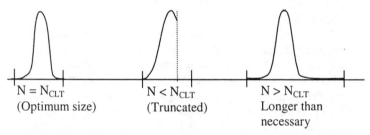

$N = N_{CLT}$	$N < N_{CLT}$	$N > N_{CLT}$
(Optimum size)	(Truncated)	Longer than necessary

Figure 4: The optimum CLT sample size (left graph) helps determine if a distribution will be truncated (middle) or longer than necessary (right)

The right sample size N for the CLT process has to be such that it allows an approximately normal distribution to develop fully. Strictly speaking, this means N must include the extreme case, where r_L shows up in all samples, within k sigma's. In practice, we know no matter how random the process is, three sigma's already include roughly 99.74% of all data, four sigma's 99.99%, and five sigma's 99.999%. We therefore will only require N to include k_0 sigma's, where k_0 is a number of our own choosing between, say, 3 and 6. The choice for k_0 depends on how precise we want our result to be.

Since $N_S=N$ for the chosen condition, equation (21) becomes:

$$N-Nr_L = k_0\{Nr_L(1- r_L)\}^{1/2} \qquad (22)$$

Solving for N and call it N_{CLT} to specify that it is the required sample size for a CLT process:

$$N_{CLT} = k_0^2 r_L/(1-r_L) \qquad (23)$$

Processes with sample size N smaller than N_{CLT} will not contain enough sigma's (on at least one side of the curve) to qualify as normally distributed. Processes with sample size larger than N_{CLT} are "normal", but they may not contain much more information than the distribution with length of exactly N_{CLT}.

Condition (23) explains why in most practical cases, a number between 10 and 30 is sufficient for the Central Limit Theorem to work, and why the required sample size increases with the skewness of the original distribution. These results, which are known to all statisticians, cannot be deduced from the (textbook) Fourier transform proof for the Central Limit theorem.

Processes satisfying N_{CLT} are "optimized processes" and their distributions "optimized distributions". The concept of "optimized distribution" is helpful in determining the optimum sample size N for complex experiments. For example, for multiple-coin tossing experiments by a human being the estimated parameters are $r_L = r_S = 0.5$, $k_0 = 3$ (three sigma's). The optimum size is:

$N_{CLT} = 3^2(0.5/0.5) = 9$

Thus, the optimum sample size is N=9 for each trial. The distribution is not fully developed into a normal distribution (i.e., truncated) when N <9. No new information can be obtained with N >9.

However, if we simulate the same experiment with a super computer we may want to choose $k_0 > 3$. That is because the k factor for a typical computer is much larger than 3; and although the amount of information outside three sigma's is very small, they could be of interest in specialized projects.

Trial size requirement for the Central Limit Theorem

The goal of performing multiple trials is to establish a distribution. If the number of trials is insufficient, the distribution is not well developed and may lead to wrong or inaccurate interpretations.

Since we are concerned mainly with the Central Limit Theorem averaging process, which results in binomial distributions, we will only analyze the case of binomial distributions.

A distribution is considered well developed when its sigma value has stabilized. Operationally this means the occurrence of a couple successive maximum deviations will not alter sigma significantly.

Let's start when the trial size is M. At this moment the calculated value for sigma σ_M is:

$$\sigma_M = (S_M)^{1/2} \approx \{M\sigma_M^2/(M-1)\}^{1/2} \qquad (24)$$

Where S_M is the standard variance after M trials. Since M is much larger than 1, we will use the trial size M for both numerator and denominator from here on for simplicity.

Next we assume the unlikely event of two maximum deviations (at $k\sigma$) occurring in a row, giving an additional $2(k\sigma)^2$ to the numerator inside the bracket. For the denominator, now we have to use M+2.

$$\sigma_{M+2} = (S_{M+2})^{1/2} \qquad (25)$$

$$=\{\{M\sigma_M^2 + 2(k\sigma)^2\}/(M+2)\}^{1/2} \tag{26}$$
$$=\{(M/(M+2))\sigma_M^2 + 2k^2\sigma^2/(M+2)\}^{1/2} \tag{27}$$

Since the change in sigma is insignificant for the condition of interest, we will simply say $\sigma = \sigma_M$ and pull it out of the square root bracket:

$$=\{M/(M+2)\}^{1/2}\sigma (1+ 2k^2/M)^{1/2} \tag{28}$$

Using the approximations $(1+x)^{1/2} \approx 1+ x/2$ for $x \ll 1$, and $M \approx M+2$:

$$\approx \sigma\{1+ k^2/M\}= \sigma + \sigma k^2/M \tag{29}$$

So while the average value converges with increasing sample size, sigma converges with increasing number of trials.

For σ to be stable, the term $\sigma k^2/M$ cannot be greater than a given fraction s of σ. This leads to the following condition for a stable σ:

$$\sigma k^2/M \le s\sigma \tag{30}$$

Solving for M:

$$M \ge (1/s)k^2 \tag{31}$$

Since it takes time to run a trial the number of trials M tends to be small (whereas the sample size N could be very large.) Because of this practical limit, a realistic value for s is believed to be at least several percentage points in most distributive systems (say, after two hundred trials.) Let's be liberal and put this tolerance at 10%. We obtain the following condition for the minimum trial size for a CLT process:

$$M \ge 10k^2 \tag{32}$$

This time we do not have the option to choose k, because it is a measure of the randomness level of the process, which affects the convergence of sigma.

The "rule of thumb" of 201 trials currently used by many statisticians corresponds to a k value of 4.48, which is very reasonable and should work for most cases.

For coin tossing by human beings, k^2 is believed to be between 5 and 9. This means only 50 to 90 trials are needed to establish a clear pattern.

NOTE: The ratio r_L/r_S was present in the condition for the sample size, but not for the trial size of CLT processes. That is because we assume that all trials are for a CLT distribution which is supposed to be well developed, symmetrical and therefore has k sigma's available on both sides of it.

The generalized binomial theorem and Central Limit Theorem

We will summarize our finding so far in two theorems, which are improved versions of the existing binomial theorem and Central Limit Theorem:

GENERALIZED BINOMIAL THEOREM

Let p be the inherent propensity of an arbitrary distributive entity X, simple or complex. Further let N be a sample size that satisfies:

$N \geq k_0^2 r_L / r_S$

Where r_L and r_S are the larger and smaller values of the pair {p, (1-p)} respectively, and k_0 a chosen number greater than 3.

Then multiple trials of sample size N will form an approximately normal distribution with the following parameters:

Mean, $\mu = Np$

Sigma, $\sigma = \{Np(1-p)\}^{1/2}$

In order to guarantee that the distribution is well developed, the number of trials M has to satisfy the following relationship with the k-factor:

$M \geq k^2 / s$

Where s is a low enough fractional uncertainty in the calculated sigma value, usually taken to be 0.1 (i.e., 10% uncertainty).

By modifying the generalized binomial theorem slightly, we get the generalized Central Limit Theorem:

GENERALIZED CENTRAL LIMIT THEOREM (*Distribution theorem*)

Let μ be the mean value of an arbitrary distribution with standard deviation σ_0, minimum value a, and maximum value b. Further let N be a sample size that satisfies:

$N \geq k_0^2 r_L / r_S$

Where r_L and r_S are the larger and smaller values of the pair (μ - a)/(b-a) and (b-μ)/(b-a) respectively, and k_0 a chosen number greater than 3.

Then multiple trials of the average value of sample size N will form an approximately normal distribution with the following parameters:

Mean, μ

Sigma, $\sigma = \sigma_0 / N^{1/2}$

In order to guarantee that the distribution is well developed, the number of trials M has to satisfy the following relationship with the k-factor:

$M \geq k^2 / s$

Where s is a low enough fractional uncertainty in the calculated sigma value, usually taken to be 0.1 (i.e., 10% uncertainty).

The alert reader will notice that the generalized binomial theorem is a special case of the generalized Central Limit Theorem where the starting distribution is the simplest distribution possible (only 2 elements with inherent propensities p and 1-p), and the resulting samples are total values instead of averages. The connection between the two theorems is not at all surprising, as both are natural consequences of the equal opportunity principle.

Since the generalized Central Limit Theorem is applicable to all distributions, the preferred name for it is the "Distribution theorem". Note that we made improvements over the traditional binomial theorem as well as the traditional Central Limit Theorem by adding conditions for the sample size N and the trial size M.

C. SUCCESSIVE CLT PROCESSES

Starting with an arbitrary original distribution with mean value μ_0 and standard deviation σ_0, by taking N samples at a time and average them we get the CLT distribution, which is a binomial distribution with the following parameters:

$$\mu_C = \mu_0 \tag{33}$$
$$\sigma_C = \sigma_0/N^{1/2} \tag{34}$$

For convenience we will force the mean of the CLT distribution to be zero, and divide its sigma by its true mean μ_0. The first action is permissible because it is simply a translation move along the horizontal axis. The second action reduces the scale of the horizontal axis by the factor $(1/\mu_0)$. The results are:

$$\mu_1 = 0 \tag{35}$$
$$\sigma_1 = (\sigma_C/\mu_0) = \sigma_0/(\mu_0 N^{1/2}) \tag{36}$$

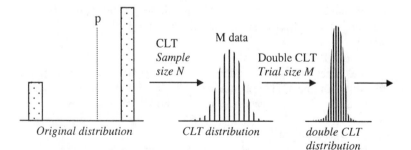

Figure 5: The Central Limit Theorem averaging process does not stop after one operation. In reality, it goes on and on in multiple stages. While we are holding this book, there are all kinds of double, triple, quadruple CLT processes going on all over the universe.

To achieve the above, each sample of size N from the original distribution will first be averaged to give us a value D_0, then operated on by the formula:

$D_1 = (D_0-\mu_0)/\mu_0$ (37)

Since the average of all D_0's is μ_0, the average of all D_N's, which is μ_N, is also zero. Thus (34) is achieved. Since all data points are divided by μ_0, the resulting sigma is also divided by μ_0, and (35) is also achieved.

Since each sample of size N gives us one data point in the CLT distribution, after M trials the total number of data points in the CLT distribution is also M.

Figure 5 is a faithful description of how we have obtained data in chapter 2. From the individual propensity of 0.5 for both heads and tails we established the propensities for various coin combinations. Each of these propensities was called p. Thus, each original distribution was a highly fragmented binomial distribution with only 2 possibilities (left picture of figure 5). We then constructed the CLT distributions (center picture of figure 5) with various sample size N by repeating some combinations 144 times (i.e., M=144), others at 200 times, 201 times, 220 times, and 240 times. Finally we calculated sigma's from raw data.

Now think about this: The CLT distribution itself can be considered as an original distribution and the M trials as a sample of size M for a new CLT process, which we will call "double CLT". This successive CLT process could go on to triple-CLT, then quadruple-CLT, etc. ad infinitum.

We will therefore push the Distribution theory one step further. We will consider each CLT distribution as one trial in the process of forming the double-CLT distribution (right picture in figure 5). Let M be the number of trials in each CLT distribution, σ_1 its sigma; then the resulting double-CLT distribution will be of sample size M and σ_2, where:

$\sigma_2=\sigma_0/M^{1/2}$ (38)

And since the same random number generator was used in the formation of the CTL and double-CLT distributions, the k-factor is unchanged. The convergence limit for the double-CLT distributions will be:

$\varepsilon_2 =k\sigma_2= k\sigma_0/M^{1/2}$ (39)

We will take all M trials in a CLT distribution and average them. This average is one data point D_2 in the resulting double-CLT distribution. All D_2 must meet the condition:

$|D_2| \leq \varepsilon_2 =k\sigma_2= k\sigma_0/M^{1/2}$ (40)

It is more convenient to measure D_2 in terms of sigma's, giving the alternative condition:

$$|D_2|/\sigma_2 = |D_2|M^{1/2}/\sigma_1 \leq k \qquad (41)$$

This means all D_2's have to be within +/-k number of σ_2's. In table 1, D_2 and D_2/σ_2 are listed in the two extreme right columns. Of special interest is the last column (on the right), where D_2 is expressed in number of sigma's.

There are 8 cases between 1 and 2 sigma's, and two cases over two sigma's out of 35 total data points. The k factors for the number generators used in this experiments are definitely larger than 4. So it is confirmed that all data are within the pre-established k-factor ranges.

TABLE 1: Double-CLT average test results
(Test date: August 2001)

Combination	Sample size	Trial size	CLT sigma	Double-CLT Average of M trials	Double-CLT Average Of M trials (in number
	N	M	σ_1	D_2	of σ_2)
1H (1)	10	144	31.00%	2.00%	0.77
1H (1)* - L	*10*	*240*	*31.33%*	*1.92%*	*0.95*
1H (1)	20	144	23.00%	-2.00%	-1.04
1H (1)* - L	*20*	*240*	*22.00%*	*0.29%*	*0.20*
1H (1)	200	144	7.00%	-0.14%	-0.24
1H (1)* - L	*200*	*240*	*7.00%*	*-0.96%*	**-2.12**
1H (1)	4000	200	1.48%	0.20%	1.91
1H (1)*	*4000*	*200*	*1.55%*	*-0.12%*	*-1.10*
1H (1)	40000	200	0.50%	0.03%	0.85
1H (1)*	*40000*	*200*	*0.49%*	*-0.01%*	*-0.29*
1H (1)	120000	201	0.30%	-0.01%	-0.47
1H (1)*	*120000*	*201*	*0.27%*	*-0.02%*	*-1.04*
2H (2)	1500	220	4.10%	-0.21%	-0.76
2T (2)	1500	220	4.73%	0.07%	0.22
1H+1T (2)	1500	220	2.50%	0.07%	0.42
1H+1T (2)*	*1500*	*220*	*2.42%*	*-0.01%*	*-0.06*
3H (3)	2600	240	4.76%	0.20%	0.65
3T (3)	2600	240	5.56%	0.41%	1.14

2H + 1T (3)	2600	240	2.70%	-0.08%	-0.46
1H + 2T (3)	2600	240	2.72%	-0.12%	-0.68
4H (4)	2400	240	8.63%	0.70%	1.26
4T (4)	2400	240	7.70%	0.60%	1.21
3H+1T (4)	2400	240	3.33%	-0.10%	-0.46
1H+3T (4)	2400	240	3.77%	0.40%	1.65
2H+2T (4)	2400	240	2.57%	-0.40%	**-2.41**
2H+2T (4)*	2400	240	2.70%	0.10%	0.57
2H (2)	144,000	201	0.45%	-0.02%	-0.63
3H (3)	96,000	201	0.86%	-0.02%	-0.33
4H (4)	72,000	201	1.37%	0.09%	0.93
5H (5)	57,000	201	2.14%	0.04%	0.26
6H (6)	48,000	201	3.43%	0.10%	0.41
7H (7)	39,000	201	5.19%	0.26%	0.71
8H (8)	36,000	201	8.11%	0.04%	0.07
9H (9)	30,000	201	12.73%	0.50%	0.56
10H (10)	27,000	201	19.16%	0.82%	0.61

Note 1: The asterisk (*) denotes repeated experiments for comparison
Note 2: Experiments with "L" was on Lotus 1-2-3, the rest on Excel

More interesting are the following data from the double-CLT distribution, which should develop into an approximately normal curve with enough data points. Although the total data points are quite small (35), the agreement is already excellent:

TABLE 2: Statistical data for double-CLT distribution

Parameters	Theory	Actual data
1 sigma range	68.3% of data	(35-10)/35=71.4%
2 sigma range	95.4% of data	(35-2)/35= 94.3%
Mean value	0.000 sigma's	0.093 sigma's
Sigma	1.00 sigma	0.97 sigma

These results reveal the dominant role of the CLT averaging process in all distributive phenomena. Each averaging process is itself waiting to be averaged. While the readers are holding this book, there are countless CLT processes running in the world at various levels, we are just not aware of them.

We conclude that continuous averaging is the working mechanism of all distributive processes and the Central Limit Theorem is its operational rule./

APPENDIX: MORE NOTES ON BINOMIAL SIGMA

Binomial sigma for small sample size

Recall that the formula for binomial sigma is $\{Np(1-p)\}^{1/2}$.

In chapter 2 we have seen that the theoretical sigma values for the binomial distribution match experimental results quite closely for N=10 and larger. This may give the feeling that the theory does not work for N below 10. To show that this is not the case N=1, 2, 4, 6 were tried for p=0.5, 0.6, 0.7, 0.8. Note that it is not necessary to perform the same tests for p=0.1, 0.2, 0.3, 0.4 because sigma values are the same for p and (1-p). All values listed are "normalized" sigma (i.e., sigma divided by the mean value.)

Just like cases for N=10 and larger, agreement within 10% between theoretical value and calculated values are achieved for all cases after 288 trials. There are two cases with error above 10% for 192 trials (p=0.9, N=1, 6); four cases for 96 trials (p=0.8, N=1, 2 and p=0.9 N=1, 4). These results confirm that $\{Np(1-p)\}^{1/2}$ is indeed a general formula for sigma for the binomial distribution. They also point out the need to determine the required number of trials to ensure that a distribution has been fully developed. Using the formula:

$M \geq 10k^2$

With k established as 5.73 in an earlier chapter, the condition is:

$M \geq 328$

The reader can verify that this criterion work for all cases listed in table A.

TABLE A: Experimental sigma values at various M vs. theory

P	N	Sigma, theory vs. various number of trials M				
		Theory	96	192	288	384
0.5	1	1.000	1.003	1.002	1.001	1.001
			0.3%	0.2%	0.1%	0.1%
0.5	2	0.707	0.706	0.716	0.708	0.689
			-0.2%	1.3%	0.1%	-2.6%
0.5	4	0.500	0.538	0.523	0.515	0.507
			7.6%	4.6%	3.0%	1.4%
0.5	6	0.408	0.374	0.419	0.414	0.427
			-8.4%	2.6%	1.4%	4.6%
0.6	1	0.816	0.808	0.808	0.796	0.808
			-1.0%	-1.0%	-2.5%	-1.0%

0.6	2	0.577	0.620	0.560	0.591	0.576
			7.4%	-3.0%	2.4%	-0.2%
0.6	4	0.408	0.425	0.423	0.409	0.418
			4.1%	3.6%	0.2%	2.4%
0.6	6	0.333	0.350	0.337	0.326	0.323
			5.0%	1.1%	-2.2%	-3.1%
0.7	1	0.655	0.678	0.637	0.651	0.664
			3.6%	-2.7%	-0.6%	1.4%
0.7	2	0.463	0.476	0.488	0.484	0.483
			2.8%	5.4%	4.6%	4.3%
0.7	4	0.327	0.345	0.351	0.340	0.338
			5.4%	7.2%	3.9%	3.3%
0.7	6	0.267	0.278	0.273	0.267	0.263
			4.0%	2.1%	-0.1%	-1.6%
0.8	1	0.500	0.400	0.455	0.478	0.508
			-20.0%	-9.0%	-4.4%	1.6%
0.8	2	0.354	0.391	0.365	0.360	0.353
			10.6%	3.2%	1.8%	-0.2%
0.8	4	0.250	0.244	0.268	0.254	0.257
			-2.4%	7.2%	1.6%	2.8%
0.8	6	0.204	0.191	0.204	0.211	0.202
			-6.4%	-0.1%	3.4%	-1.0%
0.9	1	0.333	0.248	0.289	0.313	0.336
			-25.6%	-13.3%	-6.1%	0.8%
0.9	2	0.236	0.246	0.254	0.257	0.253
			4.4%	7.8%	9.0%	7.3%
0.9	4	0.167	0.149	0.163	0.161	0.162
			-10.6%	-2.2%	-3.4%	-2.8%
0.9	6	0.136	0.148	0.155	0.148	0.144
			8.8%	13.9%	8.8%	5.8%

The Probability mistake of "Expected Deviation"

According to the Probability theory, the deviation of heads or tails for an unbiased coin after N trials has the "expectation value":

$(D_N)^2 = N$ (1a)

The reader may recognize this as the geometrical averaging of N, as we have established early in this chapter.

PROOF GIVEN BY THE PROBABILITY THEORY

We will use $+1$ to represent heads, and -1 to represent tails.

Equation (1a) certainly holds true for N=1 because the deviation is either $+1$ or -1, giving $(D_1)^2 = 1$. In addition, D_N can be expressed in terms of D_{N-1}, giving two possibilities:

Possibility 1: $D_N = D_{N-1} + 1$ (2a)

Squaring both sides of (2a):

$(D_N)^2 = (D_{N-1})^2 + 2D_{N-1} + 1$ (3a)

Possibility 2: $D_N = D_{N-1} - 1$ (4a)

Squaring both sides of (4a):

$(D_N)^2 = (D_{N-1})^2 - 2D_{N-1} + 1$ (5a)

Since all events are independent, the Probability theory assumes that each of (3a) and (5a) is true half of the time. With this assumption, one is allowed to take the average of (3a) and (5a) to get:

$(D_N)^2 = (D_{N-1})^2 + 1$ (6a)

Since it was proven that $(D_N)^2 = N$ for the case of N=1, by the method of induction it follows from (6a) that the following is true for all N's:

$(D_N)^2 = N$ (7a)

FURTHER DEVELOPMENT BY THE PROBABILITY THEORY

Let's consider the case for heads. Let N_{HI} and N_{HA} be the ideal and actual count for heads in N trials of an unbiased coin; since the difference of these two quantities is the expected deviation, our first reaction is to write the relationship:

$N_{HI} - N_{HA} = +/-|D_N| = +/-N^{1/2}$ (8a)

It was noted (by the Probability theory) that a gain for heads is a loss for tails and vice versa. The right side therefore has to be divided by 2:

$N_{HI} - N_{HA} = +/-(1/2)N^{1/2}$ (9a)

Noting that N_{HI}/N is the probability for heads P(H), we obtain the following after dividing (9a) by N and rearranging:

$P(H) = N_{HA}/N +/- 1/(2N^{1/2})$ (10a)

Where P(H) is the probability for heads, N_{HA}/N the actual ratio of heads (N_{HA}) over total trials (N).

Equation (10a) is sometimes used to establish the range for the probability P(H) from experimental data. We will argue that this practice is not scientifically sound. That is because equation (10a) can be rearranged to give:

$$N_{HI} - N_{HA} = +/- \{(1/2)(1/2)N\}^{1/2} \qquad (11a)$$

The term on the right side of (11a) is naturally the "expected deviation". The alert reader would recognize, however, that this "expected deviation" is exactly one sigma ($\sigma = [p(1-p)N]^{1/2}$) for an unbiased process where $p=1-p=0.5$.

In reality, one sigma is the average deviation. It is not sufficient to include all possibilities in most real life processes; and that is why the k-factor is necessary (k sigma's include all possibilities.)

Thus, "expected deviation" is another misinterpretation of reality by the Probability theory.

Written January 2001
Revised August 2001, November 2001, December 2002

126

Chapter 7

The foundation
of Distribution theory III

Distributive period and the Time Indifference Principle

We will use the finiteness of the k-factor to show that the as-is concept of probability is in conflict with reality and has to be discarded. However, we will predict the return of probability in another exciting branch of science. We will have a brief discussion on how the k-factor fits in with the performance of distributive systems.

We will show that each distributive process has its own distributive period. We will show that the distributive period is the same as the "resolution period" required by the Central Limit Theorem averaging process to resolve randomness. We will discuss the case when this "resolution period" is disrupted.

We will reach a shocking result, that distributive processes are indifferent to time, meaning that they will converge with or without the cooperation of time. We will call this discovery the "time-indifference principle". This principle is a crucial missing piece of information. Its discovery will change science forever.

In closing, we will show that the only useful part of the Probability theory is its predictions of collective behavior. We will show that the same purpose can be achieved more rigorously and more correctly with the Distribution theory.

A. THE END and THE RETURN OF PROBABILITY

The k-factor and the end of Probability

At the time of this writing, all processes that obey the normal distribution are ·considered to be fully characterized by only two parameters: Mean and sigma. This came from the assumption by the Probability theory that the standard normal curve is universally applicable to all normal processes. The k-factor is not a parameter because it is infinity for the standard normal curve.

The concept of probability was necessary only because otherwise the infinite k-factor would be an unsolved problem for the probability theory. Thus, by proving that the k-factor is finite in real processes, we have unintentionally made the concept of probability irrelevant in the study of distributions.

SYMMETRY AND THE END OF PROBABILITY

The Probability theory says that, for all systems and for all binomial processes, the probability for an event to be outside 3 sigma's is 0.0026. In reality, a process with k < 3 will never have an event outside 3 sigma's. In the language of the Probability theory, we should say that the probability for an event in this process to be outside 3 sigma's is identically zero. Thus, the concept of probability is not just controversial, it is incompatible with real distributions (see figure 1).

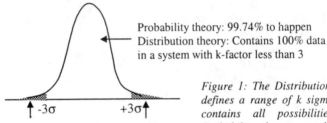

Probability theory: 99.74% to happen
Distribution theory: Contains 100% data
in a system with k-factor less than 3

Probability theory: 0.26% to happen
Distribution theory: Cannot happen
in a system with k-factor less than 3

Figure 1: The Distribution theory defines a range of k sigma's that contains all possibilities. The Probability theory, on the other hand, insists that only probabilities can be determined.

It is impossible to disprove an unverifiable theory; and that applies to the case of the Probability theory. The best we can do is to provide ample evidence that supports its antithesis, namely the Distribution theory. The Distribution theory asserts that, due to the reality of limited randomness capacity, the k-factor has to be finite for all operating systems. Experienced users of high tech machines know this from their experience. Every well-maintained machine has a characteristic range of sigma's. Some (excellent) machines never exceed 3 sigma's, some machines exceed 3 sigma's more or less frequently, some even wander out to 4 sigma's and beyond. Each machine seems to have a "machinality" outside mean and sigma, and definitely some machines are more "random" than others (i.e., having higher k-factor).

Some probability proponents may still argue that these facts still do not disprove the concept of probability, which only says something "may happen", never something "must happen". This is where scientific standards could help the jury. The choices are: An unverifiable Probability theory that has an answer for everything but cannot be held accountable for anything, or a verifiable Distribution theory that not only can predict but also is fully verifiable? The writer believes the verdict is simple: While the Distribution theory still has to stand the test of time, the Probability theory is dead!

It should be made clear that the death of the Probability theory does not means everything in Probability textbooks should be thrown away.

Quite the contrary, everything except the concept of probability is alive and well. The transformation from the Probability theory to the Distribution theory should be very smooth, because once we realize that a new interpretation is in order all the formulas will simply take a new and more credible meaning. They no longer imply the unrealistic world of probabilities, but the realistic world of converging distributions. This is a step forward, not a step backward in science.

The return of probability in THE LAST SCIENCE

The concept of probability will have to go with the probability theory to make room for the Distribution theory; but interestingly, this is not the absolute end of it. The founders of the Probability theory were great thinkers. Many mistakes of the great are treasured misplaced. Such is the case of the concept of probability.

Probability represents the limit of absolute knowledge. It came too early in the Probability theory like the boy who cried wolves. There is no place for probability in the world of distributions, which is still ruled by determinism.

However, there will be a day when science has to face the reality that it finally has reached its limits, and no more deterministic prediction is possible. This will be the same day that probability makes it eventual return. The return of probability will be dramatic and may even provoke opposition; but this time it will prevail as an integral part of THE LAST SCIENCE. The writer hopes to present the case for this last science in a future opportunity.

B. THE k-FACTOR and SYSTEM PERFORMANCE

Effects of k-factor and sigma on system performance

With all possibilities being confined within k sigma's, the normal curve for a real process will have to be different from the standard normal curve. The process of adjusting an idealistic distribution to fit reality is called "renormalization". To renormalize the normal curve we will use the relationship for the maximum deviation, which we developed in chapter 3:

$$\varepsilon_P = |N_S - Np|_{max} = k\sigma \qquad (1)$$

Equation (1) shows that 2 systems with the same sigma could have different k-factor. Obviously small k and/or sigma is desired in most man-made systems, where predictability translate to higher system price and profit. The notable exception is random number generators, where the goal is to produce randomness.

Sigma is related to the sample size N and the propensity p by the now familiar equation:

$$\sigma = \{Np(1-p)\}^{1/2} \tag{2}$$

N can be associated with factors that are sources of variations; which explains why a simpler system (low N) tends to be more reliable. In order for a system with higher N to keep sigma at a reasonable value, it is necessary to shift p and (1-p) further away from each other to reduce the product p(1-p). This shift is a move toward determinism; which explains why such a strategy is effective.

The k-factor has nothing to do with the source of noise. It is more like a "noise amplifier" (in bad designs) or a "noise damper" (in good designs). Thus, it tends to have something to do with the structural stability of the system. High but stable k-factor points to design flaw, variable k-factor points to unstable structure.

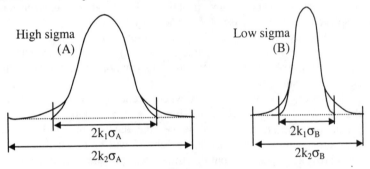

Figure 2: Both sigma and the k-factor affect system performance. High sigma and high k (case "$2k_2\sigma_A$") is definitely bad. Low sigma and high k (case "$2k_2\sigma_B$") or low k and high sigma (case "$2k_1\sigma_A$") are intermediate cases. Low sigma and low k is best (case "$2k_1\sigma_B$"). – Note: pictures not drawn to scale.

In addition to creating large deviations intermittently, high or variable k-factor is one major cause of the so-called "mean-shift" problem, where repeated runs on the same system yield the same sigma, but different mean. This can be seen from equation (1).

For the special case of random number generators, variable k-factor implies that the algorithm fails to simulate the equal opportunity principle.

The addition of the k-factor as a system parameter

At the time of this writing, the k-factor is assumed to be infinitely large. Thus, there is no reason to consider it as a system parameter.

This is a serious mistake that needs to be corrected, at least for very high end products. Take the case of two expensive high tech machines of the same model. It is customary to assume that if both machines give the same sigma and hold the same mean within certain tolerance, then they are equal in performance. In reality, we should expect each machine to have its own k-factor. The machine with higher k-factor will intermittently perform worse than the one with lower k-factor (e.g., producing unusually bad data points commonly called "flyers".) Unequal performance with same sigma is a known fact to most engineers, but no one has bothered to look for a permanent solution. That is because of the assumption that mean and sigma are the only two parameters that count.

Since the k-factor is a system constant, it is possible in principle to design a test for it. Since the range outside 3 sigma's is of particular interest and this range contains 0.26% of data in the long run; a simulated test in excess of 10,000 data points should contain sufficient information. A feasible specification for the k-factor would be the percentage of points outside, say 3 sigma's.

C. RESOLUTION PERIOD
AND THE CENTRAL LIMIT THEOREM

Distributive period

We now know that if mean, sigma, and k-factor are the same, we have two equivalent processes; but what does "equivalence" mean to a person comparing several sets of random 0's and 1's, with pages containing thousands of numbers each? Obviously, he should have the feeling that all numbers are placed randomly, so randomly that if he mixed up the first page of a set of data with the 3rd page, he would not know the difference. In fact, we expect the test results for the mixed up data will give him the same set of mean, sigma, and k-factor.

If our researcher cuts each page in half and shuffles these half pages together to form a new set of data, then re-analyzes them; most likely he will get the same sigma and k-factor (the mean of course does not change.)

But if he keeps on with this halving and mixing process, at some point he will find that sigma or k-factor, or both, will change. He may even find that each part now has a different sigma and/or k-factor. Since the k factor is supposed to be dependent only on the operating system, he knows that the data no longer qualify as test results of a distributive process. He will be forced to conclude that the working mechanism of a distributive process requires certain minimum number of consecutive events to manifest itself.

This leads to a curious observation: Each distributive process has a finite "distributive period".

This "distributive period" has never been mentioned by the Probability theory. So what is it? Where does it come from?

The necessary existence of distributive periodicity
The existence of the distributive period means that all distributive processes are periodic with respect to the number of trials (or number of samples, depending on the point of view of the analyst.) This point was missed in the Probability theory, because under the (incorrect) assumption of infinite randomness, periodic behavior should not exist.

However, once we realize that finite randomness is a reality, periodicity is a necessary conclusion. Because without periodicity there is no guarantee that a system will behave within the maximum deviation limit imposed by the k-factor.

The meaning of distributive period
The alert reader may have guessed that the distributive period is no other than the number of trials M required in the CLT averaging process to stabilize sigma. The result we found from the last chapter was:

$$M \geq k^2/s \qquad (4)$$

where s is an error fraction less than 0.10. Thus, regardless of where the starting point is in a process, k^2/s trials later we are sure that the distribution is well developed.

Let's determine how sigma achieves its stability. We will start with the variance. After T_0 trials the value of the variance is deviated from the ideal value σ_0^2 by an amount ΔS_1:

$$S = \sigma^2 + \Delta S_1 \qquad (5)$$

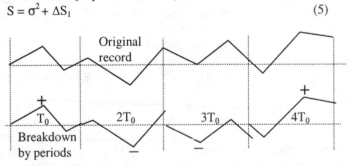

Figure 3: In data analysis, it is known that there exists a period T_0 which defines sigma. If the trial size (or sample size, depending on point of view) is smaller than T_0 the calculated sigma value will deviate significantly from the long term sigma value. If the trial size is larger than T_0 the benefits gained by increasing the trial size is significantly reduced. We call T_0 the "distributive period".

We have used the notation σ for the calculated sigma to differentiate it from the ideal sigma σ_0.

After $2T_0$ trials:

$$S = \sigma^2 + (\Delta S_1 + \Delta S_2)/2 \qquad (6)$$

In general after mT_0 trials:

$$S = \sigma^2 + \Sigma(\Delta S_i)/m \qquad (7)$$

Where the summation is taken over all ΔS_i's. Since all ΔS_i's are randomly distributed around σ_0^2, we can apply the Central Limit Theorem to the second term on the right hand side to get:

$$S = \sigma^2 +/- \Delta S_0/m^{1/2} \qquad (8)$$

Where ΔS_0 is a characteristic "variance deviation". We will not talk more about it because it is only a mathematical artifact to be replaced later.

Sigma is obtained by taking the square root of S:

$$\sigma_0 = S^{1/2} = \sigma\{1+/- \Delta S_0/(\sigma^2 m^{1/2})\}^{1/2} \qquad (9)$$

Since the second term in the bracket should be much less than 1, using the approximation $(1+x)^{1/2} \approx 1 + x/2$ we get:

$$\sigma_0 = \sigma \pm \Delta S_0/(2\sigma m^{1/2}) \qquad (10)$$
$$= \sigma \pm \Delta S_0(T/M)^{1/2}/(2\sigma) \qquad (11)$$

For simplicity, we will define the characteristic sigma fluctuation $\Delta\sigma_0$, which has to be established by experimental means:

$$\Delta\sigma_0 = \Delta S_0/(2\sigma) \qquad (12)$$

Thanks to the equivalent nature of the "$\pm$" sign, we can switch σ_0 and σ for easier identification. The expression for sigma becomes:

$$\sigma = \sigma_0 \pm \Delta\sigma_0 (T_0/M)^{1/2} \qquad (13)$$

We have left the period T_0 in (13) to remind ourselves that the calculated value for sigma is valid only when the number of trials M exceeds the period T_0, and tends to converge with respect to M/T_0 instead of M, meaning that it takes an increment in M equal to or larger than T_0 to guarantee that sigma moves in the right direction toward its convergence limit. The distributive period T_0, then, serves as the "convergence unit" for sigma.

Note that there is no harm in choosing a period T larger than T_0, as long as $\Delta\sigma_0$ is adjusted accordingly. However, if the chosen period T is smaller than T_0, it will take more than one period for sigma to stabilize.

Resolution period for the CLT averaging process

We know that the CLT averaging process is in operation in every event, working within the limit of non-randomness allowed by the system. Thanks to the continuous action of the CLT averaging process, distributions converge. Distributive period, then, is the result of a compromise between the CLT averaging process and the randomness of the system. While the CLT averaging process tries to bring the distribution to convergence at quickly as possible, system randomness delays convergence by creating fluctuations. Each period is a minimum series of events required by the CLT averaging process to achieve reasonable "local convergence" against randomness. This can be seen by the formula for the period $T_0 = k^2/s$. A larger k means a more random system, and the longer it takes for the CLT process to resolve the randomness that it is facing.

The distributive period T_0 is therefore also the "resolution period" of the CLT averaging process.

Unit distribution and mixing length

Because of the difference between p and 1-p, the practical range of the maximum deviation of a distributive process, $2k\sigma$, is less than the longest series N_{max} where the larger propensity r_L shows up in all single events. Since a distribution cannot deviate outside its maximum deviation range, all of its useful information is contained within $2k\sigma$. We will therefore call $2k\sigma$ the "unit size", and the meaningful part of the distribution within $2k\sigma$ the "unit distribution".

Using figure 4 we can set up the following relationship:
$$2k\sigma = 2k\{N_{max}\, r_L(1-r_L)\}^{1/2} = 2k^2\, r_L \qquad (14)$$
The unit size is defined as:
$$N_U = 2k\sigma = 2k^2 r_L \qquad (15)$$

Unit size is useful in data analysis because k exceeds 3 sigma's in most real processes, and roughly 99.74% of data are contained within 3 sigma's, 99.99% within 4 sigma's. So even in the case we do not know the true value of the k-factor, we could arbitrarily set it somewhere between 3 and 4, depending on the level of desired accuracy. So the working formula for unit size is:

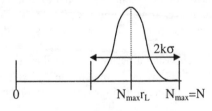

Figure 4: In our strategy to get the unit distribution, we only care about the portion of the distribution curve that contains most information. Usually it is between 3 and 4 sigma's.

Working formula: $\qquad N_U = 2k\sigma = 2k_0^2 r_L \qquad (16)$

Where k_0 is a chosen value between 3 and 4 (or higher, if more precision is required.)

We can visualize a series of event as consisting of many successive unit processes of sample size N_U. Although each unit process appears random, a clear pattern emerges when we combine a sufficient number of them.

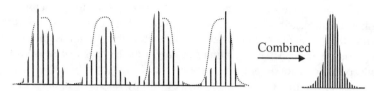

Figure 5: We can think of each unit distribution as a partial picture of the process. The mixing many unit distributions will give us the correct picture for the well-developed distribution. The minimum number of unit distributions required for this purpose is called the "mixing length".

The number of successive unit distributions that gives a well-defined combined distribution is called the mixing length L_M. Recall that the well-defined distribution is achieved with the resolution period:
$$T_0 = k^2/s \tag{17}$$
By combining (16) and (17) we get the formula for the mixing length:
$$L_M = T_0 /N_U = (k^2/s)/(2k_0{}^2 r_L) = (k/k_0)^2/(2s.r_L) \tag{18}$$
It can be seen that the mixing length increases with increasing randomness level (k-factor) and increasing deviation from determinism (Note that $r_L \geq 0.5$, with maximum deviation from determinism at 0.5)

The name "mixing length" was chosen because a distributive process is, in a sense, a mixing of many random elements to yield a certain level of uniformity in the form of the normal distribution.

D. THE TIME-INDIFFERENCE PRINCIPLE

The role of time in distributive processes

The Probability theory has mistakenly assumed that distributive processes have infinite randomness because it failed to realize that every process has to be performed by a system. No matter how potentially random a process is, the non-randomness of the system is the controlling factor. This mistake of the Probability theory can be seen right from the approximated formula for sigma and convergence limit for a binomial process:
$$\sigma = \{Np(1-p)\}^{1/2} \tag{19}$$
$$\varepsilon_P = k\{(1-p)/(Np)\}^{1/2} \tag{20}$$
Since k is infinity, if N is infinity σ will have to be infinitely large, and ε_P either infinitely large or inconclusive. It is non-sense to even talk about a standardized normal distribution that have infinite large

standard deviation and may not converge. It is mind boggling to think that the Probability theory could be around for so long with such serious paradoxes unsolved.

We have solved these paradoxes by realizing that the k-factor is finite for all distributive systems. This means there can be no process with infinite sample size (or trial size for that matter.) In hindsight, this point is obvious. After all, infinity is only a mathematical idea whose sole purpose is to simplify our treatments of physical processes. It is our task to determine whether the idea of infinity should be applied in a given case. In the case of distributive processes, the Probability theory has applied the idea of infinity incorrectly, with disastrous results.

The problem of infinity led the Probability theory to a situation where it was doomed to be a pseudo-science:

Unknowns (3): Infinity, single events, and collective distributions.

Known (1): Statistical theory on distribution (some form of binomial theorem or Law of Large Number, or Central Limit Theorem, etc.)

The probability proponents would argue that the concept of probability is another known parameter. In reality, the quantification of this concept is drawn from statistical theory, which has been already counted elsewhere. Probability therefore cannot be considered as an independent parameter.

By simple accounting the number of residual problems is equal to the unknowns minus the known. It is no surprise, then, that the Probability theory has two residual problems:

1. Event probability is not verifiable.
2. Convergence of distributions cannot be proven (or proven by causing conflict with probability as in the case of the Law of Large Number).

The Distribution theory outperforms the Probability theory not by doing more, but less. First, it distances itself from single event probabilities, realizing that they are outside the scope of any theory of collective distributions. Second, it removes the infinity problem by recognizing that infinite randomness does not exist. The first action was important, but the breakthrough was the second. With it, distributive processes became a problem with one unknown and one known, thus scientifically solvable.

With the problem of distributions formally solved, a surprising result emerges. To appreciate this result we have to mention classical physics.

Classical physics assumes that all events are deterministic. This view is presented graphically in the left picture of figure 6. The classical assumption breaks down in distributive processes, such as the tossing of N coins, which varies within a random range.

It is customary to think of distributive processes as exceptions to the laws of physics (i.e., classical physics.) Shockingly, by looking at figure 6 we see the reverse: Classical physics is only a very special case of Nature!!!

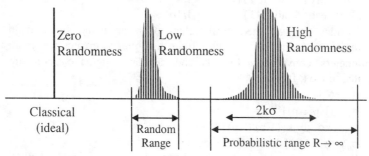

Figure 6: While the classical view (left) is inadequate because it assumes zero randomness, the Probability theory assumes also incorrectly that randomness is infinity. Reality is between the two cases, and includes them as extreme limits. The Distribution theory is consistent with this reality.

It is tempting to make the statement that distributive processes rule the world of physics, and classical physics is but the limiting case of distributive processes where the random range can be approximated as zero. This, however, is not strictly correct. There is one important factor that differentiates the class of distributive processes that we have been studying and the processes of classical physics. That is time. The alert reader must have noticed that there has been no mentioning of time at all in the Distribution theory so far.

The ultimate process, then, is a combination of the static distributive process like we have been studying and the dynamics of time. That happens to be the realm of quantum physics (and selected chaotic processes.)

But it is too early to discuss quantum physics. Let's return to our distributions. Do they really have nothing to do with time? Does time affect them? We will discuss these questions in the next few sections.

The near-perfect convergence of distributions

A distribution may consist of many combinations, each with its own propensity p and its own distribution. This proliferation of distributions is best illustrated with an example. Take the case of tossing 4 unbiased coins at one time. The possible combinations and inherent propensities are:

TABLE 1: Combinations in 4-coin tosses

Combination	Inherent propensity p
4 heads (4H)	1/16
3 heads 1 tails (3H1T)	1/4
2 heads 2 tails (2H2T)	3/8
1 tails 1 heads (3T1H)	1/4
0 heads 4 tails (4T)	1/16

Besides the two theoretical values of mean μ and sigma σ in binomial processes, the reality of finite randomness creates two more parameters, convergence limit ε_P and calculated sigma σ_0, both are related to the k-factor:

$$\mu = Np \tag{21}$$
$$\sigma = \{Np(1-p)\}^{1/2} \tag{22}$$
$$\varepsilon_P = k\sigma/\mu = k\{(1-p)/(Np)\}^{1/2} \tag{23}$$
$$\sigma_0 = \sigma \pm \Delta\sigma_0 \{(k^2/s)/M\}^{1/2} = \sigma \pm k(\Delta\sigma_0)/(Ms)^{1/2} \tag{24}$$

If we toss 4 coins N times and record the occurrences of all 5 possible combinations, we will see convergence to the ratios listed in table 1. Let's take the 4 tails combination and study it in detail.

Note that the convergence limit ε_P is dependent on the sample size N, though more correctly we should say ε_P is dependent on N/T_0, knowing now that T_0 is the resolution period of the equal opportunity principle.

The reality of finite randomness dictates that the process will evolve as multiple unit processes with fixed sample size.

Recall that the total capacity of a system is $U=K/P_{min}$, where P_{min} is the smallest propensity that is allowed to occur. We will define $N_0=1/P_{min}$ as the random capacity because it limits the propensities that can and cannot occur.

When the random capacity N_0 is reached instead of getting the perfect mean value $\mu = (1/16)N_0$ for 4T, we are only guaranteed to be somewhere in the range $\mu \pm k\sigma$.

But if we keep increasing N to many multiples of N_0, an interesting effect takes place. The Central Limit Theorem averaging process will start acting on units of N/N_0 as if it is the sample size, giving us $\mu +/- k\sigma/(N/N_0)^{1/2}$, which would converges if N could approach infinity. However, since the total capacity U itself is limited, as some point the convergence will stop. Thus, there exists a "zero-point fluctuation" that a system can never get out of. "Zero point fluctuations" are known to exist in all man-made systems. Our point is that it also has to exist in Nature. However, the "zero-point fluctuation" should be negligibly small when compared to the scale of the distribution, and is of interest only in very specialized cases.

Next, we apply a similar analysis to the CLT distribution of an individual propensity.

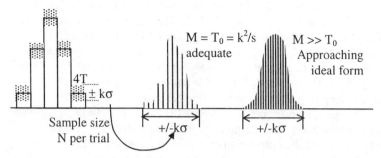

Figure 7: The two CLT distributions (middle and right) are not long-term descriptions of all combinations in the 4-coin toss, it only describes the behavior of the 4 tails combination. Note that the inherent propensity for the 4 tails combination is p=0.0625, meaning that it will be only about 6.25% of the long-term distribution when we tally all 5 possibilities with 4 coins. However, its own behavior must also obey the CLT averaging process. When the number of trials M reaches the distributive period T_0 the CLT curve for 4 tails contains adequate information. When M is much larger than T_0 it approaches the ideal CLT form.

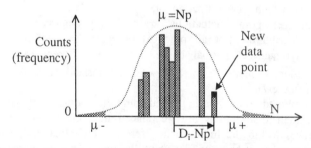

Figure 8: To add a new data point to a distribution, we look for the horizontal location that matches its value. Then we add 1 to the count at that location (in the vertical direction). The final distribution curve is the approximately normal curve that best fits all data.

Since the propensity for 4 tails is 1/16, on the average we will have to toss 16 times to get one occurrence. There is no rule that says we cannot consider each toss as one trial and the number of tosses as the number of trials. (This would amount to a Central Limit Theorem averaging attempt of sample size 1, which does not meet the sample size requirement and will lead nowhere.) Obviously, the better strategy is to toss N times, record the number of occurrences of the combination that we have chosen, then keep repeating the same procedure (of N tosses) to get a total of M trials. To follow the convention established in chapter 3, we will call N the sample size, and M the trial size.

With this convention in mind, for each trial of sample size N we will get a number of occurrences D_i for 4T. The mean value for 4T is at the peak of its distribution curve, which is located at $Np=(1/16)N$, so the deviation is D_i-Np. This corresponds to one count (in the vertical direction of the curve) at location D_i-Np (on the horizontal direction of the curve). The total range for the horizontal direction is 0 to N, but all counts for p should be confined within $Np \pm k\sigma$.

The distribution curve is considered sufficiently developed when M passes the period T_0. At high enough M, all possible occurrences will be present; and the distribution curve becomes more and more refined.

Since it is impossible to reduce the range $\pm k\sigma$ to zero, we can never get a CLT distribution with zero width (see center and right pictures of figure 2). By increasing M we get closer and closer to the ideal value for sigma σ. When $M=N_0$ (or whatever random limit applicable) sigma will be in the range $\sigma \pm k(\Delta\sigma_0)/(sN_0)^{1/2}$. Again, the Central Limit theorem averaging process will operates on units of M/N_0, giving us back the exact range $\sigma \pm k(\Delta\sigma_0)/(sM)^{1/2}$, which would converge to σ when M tends to infinity as multiples of N_0. Again, because of the limit in total capacity, the system will instead reach the "zero-point fluctuations" which it cannot get out of.

But the "zero-point fluctuations" are negligibly small compared to the total distribution. For all practical purposes, we will obtain a perfectly symmetrical normal curve, which is different from the standardized only in its finite range of $\pm k\sigma$ instead of $\pm \infty$ (right picture of figure 7).

The combined conclusion is that, although individual combinations fluctuate, their distributions are guaranteed to achieve near-perfect convergence, within the limit of the "zero-point fluctuations", when either the sample size N or the trial size M becomes arbitrarily large.

Simultaneous vs. sequential events

Imagine 5000 perfect cloned copies of human beings, each tossing an unbiased coin at the same time somewhere in the vast universe to give us the first set of data, and one of the cloned human being in the group tossing the same set of coins one at a time. Will we have equivalent results?

As stated earlier, the CLT averaging process is whatever is left after randomness. Our approach, then, is to compare the random and non-random factors involved in the two experiments.

TABLE 2: Comparing coin tossing experiments by human clones

Case	Randomness	Non-randomness
5000 clones, 1 toss	No sense of sequence	Identical systems
1 clone, 5000 tosses	Fatigue, distraction	Same system

Since fatigue and distraction could be discounted as minor factors, the major difference between the two cases is that there is no sense of time sequence in the case of the 5000 clones because each tosses his coin at exactly the same time as all the others.

We have established earlier in the chapter that time sequence is important in evaluating 0's and 1's produced randomly by a computer. However, it is clear that if we run the data backward (first point as last point and last point as first) the situation is equivalent. Thus, the deciding factor in this case is time sequence, not time order. But what is time sequence? It is but a manifestation of non-randomness, which allows the Central Limit Theorem to exercise its power.

But time is not the only non-random factor available in the universe. In the case of the 5000 clones, the non-random factor is the fact that they are identical copies. Thus, we expect to see the CLT averaging process working with them, too. At the same time, it is reasonable to expect some differences between multiple clones and the single clone, because even though identical clones are equivalent to one another, the fact that they are separate entities should add some randomness to their combined action (or, think about it, it could be the other way around.)

Since the CLT averaging process only requires non-randomness to operate, the reality of finite randomness guarantees that it is always in operation in various extents depending on the left over non-randomness available to it. Thus, there are always averaging processes going on everywhere in the universe. Thanks to this averaging process, the convergence of distributions is guaranteed.

Distributive determinism

With guaranteed convergence, seemingly random events will have to somehow fall in place after a sufficient number of trials so that the distributions work out correctly.

We can compare a distribution to a jigsaw puzzle. While the exact location of the next piece is unknown, it will be a part of the eventual picture, which is deterministic and in fact can be predicted ahead of time. This is truly amazing!!!!

It is important to differentiate the "determinism" of distributive processes from that of classical processes (i.e., processes that are subjected to the laws of classical physics.)

-Classical processes: All individual events are deterministic.

-Distributive processes: Individual events may not be deterministic, but their distribution is deterministic in the sense that it will converge to pre-determined order.

We will use the term "distributive determinism" to differentiate the determinism of distributive processes from the conventional meaning applicable to classical processes.

By recognizing the existence of "distributive determinism", the Distribution theory has advanced one huge step ahead of the Probability theory. This breakthrough is essential for the understanding of quantum mechanical phenomena, as will be seen in a later chapter.

The time-indifference principle (TIP)

Return to the case of the 5000 clones tossing one unbiased coin each at exactly the same time. If we inspect the result, most likely we will find the heads/tails ratio is 0.5 within certain tolerances. This is a distribution, and it does not require time to form. It is tempting therefore to state that the convergence of distributive process is time independent! In the more normal case, however, convergence is a time process and the equal opportunity manifests itself through time. So to avoid confusion we will alter our statement and say that convergence is "time-indifferent", meaning that they will happen with or without the help of time. The fact that distributions are deterministic will be called "the time-indifference principle".

To fully appreciate the astounding implication of the time-indifference principle, let's use an example. In this example we toss 10 coins at a time and keep track of the counts of all distinguishable combinations (there are 11 of them). Intuitively it seems these 10 coins could turn up, say 6 heads and 4 tails any time they want. In reality, they somehow miraculously limit the number of occurrences of this combination so that its ratio converges in the long run. This is so incredible that some of us may wonder if the 10 coins have consciousness. Some readers are probably aware of the parallels of this example in quantum mechanics for the cases of photons and electrons. This is why many scientists and philosophers brought up the issue of consciousness in quantum mechanics. This issue is still a subject of great controversy at the time of this writing.

The explanation is simple and has nothing to do with consciousness. The miraculous behavior of distributive processes is a necessary result of the "time-indifference principle", and quantum processes are no exception.

Limits of the time-indifferent principle

Take the case of a random number generator that can only produce 16 consecutive unbiased 0's or 1's in a row (for a k factor of about 4 sigma's). Let's say the last 15 digits that it has just produced are all 0's. We decide to run the program again. Can we say that it is impossible for the first 10 digits this time to be all 0's because that would add with the last 15 digits to 25 (for a k-factor of about 5), exceeding the limit of 16?

We have established in the last chapter that the Central Limit Theorem works on whatever non-random element is left in the system. Its operation is defined by whatever in the system that has not been taken over by finite randomness. By ending a program and re-run it, the programmer has created an abrupt change in the process. This abrupt change itself must be a distributive process, but it also must have a much higher k-factor compared to the continuous running of the program. As far the CLT averaging process is concerned, this is an abrupt increase followed by a recovery of the k factor. For all practical purposes, it is safe to assume that the k-factor of the disruption process is much higher than that of the program and may effectively wipe out the limit imposed by the CLT process in the last program, forcing the CLT process to start anew when the program is rerun. Thus, the k-factor most likely will not be able to connect the end of the last program to the start of the new program and we simply have a new beginning.

The answer to the question, then, is that it is even possible for the first 16 digits to be all 0's, not just 10 digits, when the program is rerun.

This is a case where the time-indifferent principle is violated. We therefore need to keep in mind that the time-indifferent principle works only when the k-factor of the process is kept within certain reasonable limit.

Before leaving this section, let's present a milder case where the time indifference principle may not work.

Take a football chief referee. Every game he has to flip a coin at least once. Say he averages 20 games a year and has been working for 10 years; that would be 200 flips. In addition, once in a while he would flip a coin in a bet with a friend or just for fun. If we somehow could keep a sequential record of all these coin flips as our first set of data, then ask him to flip a coin continuously 200 times for the second set of data. Could we assume that the two sets will have the same set of mean, sigma, and k-factor? Mean possibly yes, sigma maybe, but k-factor? Very unlikely!

The reason the k-factor should be different for the two sets of data is simple. The k-factor is a measure of randomness level, and it is unlikely that tosses that are too far apart in time would be governed by k-factor values that are reasonably close to one another. A changing k factor means the Central Limit Theorem operates differently. The k-factor for the first set of data, collected from fragmented coin flips, may vary outside a reasonable range. In such case, the time indifference principle would not apply.

E. THE RETURN OF DETERMINISM

From our discussions, the outcome of an event in a distributive process depends on not only the inherent tendency of the object, but also the process applied to it. Based on its construction, the Probability theory is applicable only to the collective outcome all events, but not to the individual events. For this reason, the theory should not make any claim on individual "probabilities". The idea of uniform probabilities is absurd because it implies that all events are process independent; which contradicts reality.

On the other hand, the success of the Probability theory in predicting collective distribution is understandable. Take, for example, the case of radioactive decay. The Probability theory maintains that it is impossible to predict when an atom will decay, yet it predicts the collective rate of decay of a large group of atoms with amazing accuracy. This is one frustrating puzzle for many scientists, most notably Einstein, who could not accept the concept of probability.

Now we are in a position to solve this puzzle. Since all atoms in a group undergoing radioactive decay are equivalent, the combined propensity and the randomness capacity are the same for all of them. Each atom qualifies as a sample in a distribution. The decaying sample usually consists of a very large number of atoms, which satisfies the convergence condition dictated by the Central Limit Theorem. We therefore expect all ratios to converge, giving predictable collective result. The same logic explains other distributions that are successfully predicted by the Probability theory.

It should be emphasized, however, that even within its scope of application, the Probability theory is incomplete because it unknowingly assumes that the final distribution is independent of the process leading to it. In reality, a number of combinations predicted by the Probability theory are prohibited and can never occur.

We are now ready to make the following closing statements for this chapter:

1. The classical probability concepts of "single event probability" and "multiple event probability" are useless because their

numerical values are drawn from long-term distributions, which have nothing to do with event probabilities.

2. Absolute randomness does not exist. All random systems are limited in their randomness capacity and randomness level.

3. The "equal opportunity principle" balances randomness with the Central Limit Theorem averaging process to force convergence in distributive processes.

4. Collective distributions are deterministic and predictable.

5. Distribution processes obey the "time indifference principle", meaning that N simultaneous events are the same as N events spread out arbitrarily over time.

6. Not all combinations predicted by the Probability theory can be realized. Certain extreme combinations are prohibited because of the reality of finite randomness capacity.

7. The Probability theory incorrectly assumes that an infinite number of sigma's is required to include all possibilities in normal distributions. In reality, the number of sigma's (k) is finite.

8. The Probability theory incorrectly assumes the same behavior for all distributive processes. In reality, the distributive behavior is system dependent. More specifically, each system has its own characteristic k factor.

9. The name "Probability theory" is misleading. By removing the concept of probability from distributions, we have the "Distribution theory".

Written April 2001
Revised August 2001, November 2001, December 2002

Additional reading for chapter 7

Symmetry, synchronicity
and the meaning of space-time

The skeptical reader may find the treatment of time in chapter 7 still leaves many open questions. I hope that this article will address most, if not all, of them.

Inapplicability of the space-time model in distributive processes

With the dominant influence of Einsteinian physics, today when we think of time we immediately think of the speed of light. "Nothing can move faster than light" is indeed one of the best known slogans of today's science.

We should keep in mind, however, that the speed limit imposed by relativity is meant for processes involving some form of information exchanges among entities existing in space and time. I will argue that this is not the case in distributive processes.

The success of symmetry and the end of probability

It is not an exaggeration to say that symmetry is Nature's most puzzling mystery. We know that symmetry exists but we do not know why it should. To those who say that there is nothing mysterious about symmetry because it is only the counterpart of conservation laws; I would remind them that conservation laws, too, are mysterious. Using one mysterious property of Nature to dispel its mysterious counterpart will not help us solve the mystery.

So we may as well accept the fact that symmetry is mysterious. I would even go further and say that symmetry is mystical. It is mystical because it somehow recognizes the differences and similarities of events that don't relate to one another by the cause-and-effect mechanism that we have been accustomed to. Just take the familiar case of a coin toss. Because we have taken it for granted, we tend to be oblivious to the fact that the convergence of the long-term ratio to 0.5 is indeed very mystical.

Science has tried to explain away this and many other mystical phenomena with the probability theory. This strategy seemed adequate for three centuries; but with the advent of the modern computer, the probability explanation is no longer acceptable. The reason for this has been covered in depth in chapter 3, but just to refresh the reader's memory, I will repeat it here.

There are two well known theorems in computing science. The first states that it is impossible to build any machine with infinite random capacity. The second states that a system can never produce more randomness than it possesses. Taken together, these two theorems imply that even the most powerful random number generator of the distant future will repeat itself at some point. This remarkable result is not just in the mind of the mathematicians. It has been verified on many of the most powerful random number generators in existence today.

I will pretend that I have the most powerful random number generator of today at my disposal; and I already know that it will repeat itself after a series of N numbers in my coin tossing experiment (with equal probability for heads and tails). I will pay attention to the first number that this random number generator produces. Let's say this number happens to be 0. I will immediately know that the $(N+1)^{th}$ number will also have to be zero. This means for the $(N+1)^{th}$ number, the probabilities for 0 and 1 are not equal. More specifically, the probability for 0 to occur is 100% (certainly will happen), whereas the probability for 1 to occur is 0% (certainly will not happen). This is a blatant violation of the probability hypothesis, which assigns the same probability to every random number generated by the computer.

Some Probability proponents may cry "Foul!" and tell me that this is an unfair test, because the Probability theory describes Nature, and computers (i.e., random number generators) are poor imitators of Nature. However, this very objection is a partial concession, because it indirectly confirms that at least in the operation of computers, the probability hypothesis is only an approximate calculating procedure that breaks down at the limit. In other words, probability is an incomplete description of computer operation! *(Also see note 1).*

Of course, if we didn't have anything else, then the only option would be to stay with probability and accept its drawbacks. However, as I have shown in chapter 3 and 4, by replacing probability with symmetry, we can fully account for not only computer operation, but also other phenomena in distributive processes that the probability hypothesis could only provide partial or unsatisfactory explanations.

Since symmetry is already a part of the foundation of science, and the Probability theory is a source of confusion with no added value, the only logical action is to eliminate the Probability theory from science. This simple action will allow science to move forward with a much stronger foundation than ever before.

Symmetry, propensity, and synchronicity

We know that long-term propensity is conserved thanks to the principle of symmetry. In the processes of classical physics, every phenomenon satisfies symmetry. Distributive processes are more complex, because here symmetry applies only to the fictitious "average phenomenon" that represents a large enough number of phenomena.

Since we have been conditioned to think in terms of cause and effect, the natural question that we have is "What causes a set of N events to combine together in such a way that the long-term propensity is conserved on the average?"

This question is best answered with an analogy. Imagine an orchestra with hundreds of musicians performing a complex symphony by Beethoven or Mozart. What "causes" the musicians to perform in harmony? The answer is: The "synchronicity" inherent with the original music. Thanks to this synchronicity, harmony is achieved by the orchestra without any cause-and-effect relationship among the individual musicians.

Thus, the key word for individual events in distributive processes is not cause-and-effect, but *synchronicity*[2]. Long-term propensities are realized because of the synchronicity dictated by the principle of symmetry.

Before leaving this section, I would like to leave you a note regarding the history of synchronicity. Although Carl Jung and Wolfgang Pauli haven often been credited as the discoverers of synchronicity in the west, Jung himself acknowledged that synchronicity was an Asian concept, exemplified by the I-Ching[3]. To be more general, synchronicity has always been in the foundation of Indian and Chinese philosophies, which started many thousand years before the time of Jung and Pauli.

The "event dimension" of distributive processes

Let's return to the example of the orchestra, but now imagine that we record one musician at a time, then combine all the tracks together. The result is still the symphony, complete and yet without any interaction among the musicians. This is the equivalent of a typical distributive process where only one event is observed at a time. When we combine all events together we should get a coherent average without the events ever communicating with one another.

Since there is no event-to-event communication in synchronicity, the space-time restriction imposed by relativity does not apply. Thus, for all practical purposes, the synchronicity of distributive processes operates outside the realm of space and time. Since synchronicity is the manifestation of symmetry, we can state that symmetry is a spaceless and timeless property of distributive processes.

But since the individual events of distributive processes can only be recorded in space and time, we are forced to describe distributive processes in terms of space and time. The simplest description is to consider symmetry as an inherent property (or criterion if you will) already programmed in with distributive process and is therefore automatically obeyed by the set of all individual events without any communication at all.

Thus, if we insisted on describing distributive processes in terms of event-to-event communication, we would be forced to hold the position that the communication between two spatially separated events in a distributive process occurs at infinite speed! Since mathematics is no more than a specialized language, such description is mathematically permissible; but I suggest that we avoid it, because it gives the (false) feeling that distributive processes are in conflict with Relativity while in reality there is no conflict. Basically my point is: There is no room for the idea of event-to-event communication in distributive processes!

This means a new mathematical formalism is needed to describe distributive symmetry. To search for this formalism I will consider two processes: The first involves N tosses spreading over time, the second N tosses spread over space. For reference purposes, I will call the first process a time process, the second a space process.

In a time process of coin tossing, I have the option to increase the sample size in a trial by simply performing more tosses. For example, if I am not satisfied with N tosses and want a total of $N+N_1$ tosses instead, all I have to do is to perform N_1 more tosses and consider the total of $N+N_1$ tosses as a single trial. This is permissible because except for a shift in time, the experimental condition (including space) has not changed from the original N tosses to the subsequent N_1 tosses. Since the original N tosses form a sequence in time and the subsequent N_1 tosses another sequences in time, they naturally combine into a seamless sequence of $N+N_1$ tosses in time.

Looking the other way, if I labels events in a time process as 1, 2, 3, 4,…, N-2, N-1, N, and pick out an arbitrary sequence, say events 11 to 20, I must consider this event as a trial (of sample size 10) in its own right. This is true for all possible sequences within the sample size N.

Interestingly, the situation is different for a space process. To see why I will pretend that I have asked N clones at different parts of the universe to toss one coin each to get a space process of N coins, then after the experiment has been completed I realize that I actually want $N+N_1$, not N samples. Can I ask another set of N_1 clones to toss one coin each, then add the new set of N_1 samples to the original N samples and say that I have a space process with sample size $N+N_1$?

The answer is "No!" because space is not the only difference between the original N and the subsequent N_1 tosses. Time has also changed! Of course the two sets can be combined, but only as a time process. Thus, the strategy cited does not give me a single (space process) trial with sample size $N+N_1$. Instead, it gives me 2 (time process) trials, the first with sample size N, the second with sample size N_1. The bottom line is: Once a space process trial has taken place, its sample size cannot be changed!!!

This brings out a remarkable difference in time process and space process: A time process can be divided into many smaller time processes, but a space process has to be considered as a whole!

In a time process, each finite coin tossing process totaling N tosses can be considered as a part of an infinitely long process. If symmetry only applies to this infinitely long process, it would be logically permissible for all N tosses in a trial to turn up heads (or tails) without violating the total symmetry. Since this is true for all finite N's, symmetry could be violated in all practical cases, which is an absurdity.

Thus, by *reductio ad absurdum*, I conclude that, in a time process, symmetry must be satisfied by all sample sizes. Since any sequence within N samples is qualified as a sample, symmetry must be satisfied not only by the set of all events but also by all sets of consecutive events in a time process. Space process is simpler. The only requirement is that symmetry is satisfied by the set of all events.

In summary, time symmetry applies to the parts as well as the whole, space symmetry applies only to the whole.

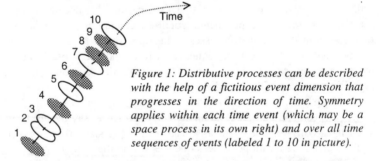

Figure 1: Distributive processes can be described with the help of a fictitious event dimension that progresses in the direction of time. Symmetry applies within each time event (which may be a space process in its own right) and over all time sequences of events (labeled 1 to 10 in picture).

This difference between space and time can be properly described by a fictitious "event dimension", of which each event is a space process. In this description, the event dimension progresses in the direction of time. Symmetry is satisfied by each event as well as all sequences of events.

Distributive process as the space-time realization of propensity

If I say "A is a good businessman" I do not necessarily mean that A has always been and always will be successful in his business endeavors. What I mean is that, if everything in A's business record is taken into account, including his most brilliant successes and his most regrettable blunders, the overall image is good.

The same meaning applies to the propensity of a process. For example, when I say "the propensity of this coin tossing process is 50/50 for heads/tails", what I mean is that if everything is taken into account, the process should give me 50/50 in heads and tails.

But what is the meaning of "everything is taken into account"? Strictly speaking, in the case of businessman A, in order to take everything into account I have to wait until he completely stops conducting business or dies before I make my judgment. In reality, if A has been in business for, say, twenty years I may have enough supporting data to make the statement "A is a good businessman" and feel quite safe about it.

The same with propensity. Strictly speaking, I will need an infinite number of coin tosses to confirm that the process gives me 50/50 in heads/tails. However, thanks to the fact that distributive processes have finite randomness, I can get away with a finite number of tosses.

There are several ways for me to confirm the propensity of a process. First, I recall from chapter 5 that the convergence limit is:

$$\varepsilon_P = k\{(1-p)/(Np)\}^{1/2} \tag{1}$$

Provided that I know the randomness level k of the process, I could achieve the desired convergence limit ε_P with a space process, say, by asking N clones to toss one coin each.

Alternatively, I can ask a single clone to toss one coin at a time N times. Formula (1) still applies, but now I must keep in mind that the condition has changed (the single clone may get tired after so many tosses and becomes more random in his tossing action, for example,) which may cause a change in the randomness level k.

Still, I could combine space process with time process by, say, asking m clones (m<N) tossing one coin each and count this as one trial, then repeat many trials to get the desired convergence limit. Again, I must keep in mind that the randomness level for this combined process may be different from the separate space process and time process cited above.

These examples show that, a distributive process is simply a realization of propensity in space or/and time. Mathematically speaking, a distributive process is a faithful mapping of the spaceless and timeless propensity to the domain of space and time via the event dimension.

Distribution theory:
The successful synthesis of determinism and randomness
At the time of this writing, there are two concurrent scientific paradigms that are diametrically opposing each other. At the one extreme is absolute determinism championed by classical physics. At the other extreme is absolute randomness advocated by the Probability theory. It is clear that a seamless combination of these two opposing paradigms is impossible.

Since the distribution theory combines both determinism and randomness, it may be the synthesis that science has been searching for. We already know that the distribution theory outperforms probability in distributive processes. Let's see how it performs in the realm of determinism.

Recall that a distributive process has two components: A space process and a time process, and the propensity can be realized in both, provided that the sample size is large enough. Now imagine a distributive process consists of many space processes with very high sample size closely packed together in the event dimensions to form a time process with very high sample size. The results are:

1. By the Law of Large Number, the propensity is realized and appears deterministic in space for every event.
2. By the Law of Large Number, the propensity is realized and appears deterministic in time for all sequences of event.

Within the accuracy of experiments properties 1 and 2 are but descriptions of a deterministic process. Thus, we conclude that the Distribution theory successfully describes all phenomena of relevance to science, including those that appear random as well as those that appear deterministic. By all likelihood, it is the correct synthesis that has been missing in science.

The amazing Time-Indifference Principle
It is well known that if a regular coin is tossed repeatedly by say a three years old child for 10,000 times, the long-term ratio will be approximately 0.5. This is of course a result of the Time-Indifference Principle. Unfortunately, coin tossing is such a familiar process that it is difficult for us to appreciate the significance of the Time Indifference Principle in it[4]. For this reason, I will now show the power of the Time-Indifference Principle by using it to explain one of the most puzzling phenomena of modern physics.

In 1803 Thomas Young gave the wave interpretation of light a solid empirical footing with his famous "double slit" experiment. By passing light through two narrow parallel slits, Young found interference patterns consisting of alternating light and dark stripes, indicating that the light coming out of the two slits acted like two interfering wave fronts.

In recent years Young's experiment has been repeated, but with one photon at a time. The result of multiple trials gave the same alternating light and dark patterns as the old experiment! This is the "single photon double-slit experiment".

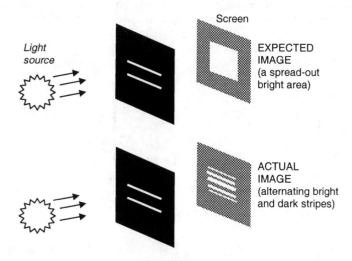

Figure 2: Young's double-slit experiment

To fully appreciate the implication of the single photon experiment, let's study the pictures in figure 2 in more detail. They are illustrations of the following procedure:

1. Cover the bottom slit and release one photon at a time toward the top slit. Each photon that passes through the slit will leave a bright spot on the screen. After many trials, the top part of the screen is covered with bright spots, as expected. (The spreading out of light in the Y direction is a known effect called "diffraction".)
2. Cover the top slit and release one photon at a time toward the bottom slit. After many trials, the bottom part of the screen is covered with bright spots, as expected.
3. Open both slits and release one photon at a time in the general direction of the two slits. It is expected that the screen is flooded with bright spots after many trials, because it is the sum of cases 1 and 2 with some photons passing through the top slit, some through the bottom slit. Instead, we get an alternating pattern of dark and bright stripes.

The result of step 3 is the same as the original Young's experiment; the big difference here is that only one photon is released each time. The alternating pattern in Young's original experiment has been explained by the wave model as the interference pattern of the two light streams coming out of the two slits, each acting as a wavefront. Obviously, with only one photon released at a time this argument collapses. After all, a single photon should go through only one slit, and its wavefront, if existed, would have no other wavefront to interfere with.

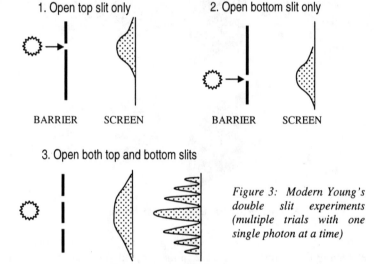

Figure 3: Modern Young's double slit experiments (multiple trials with one single photon at a time)

To get out of this dilemma, it was argued that the single photon possesses not one, but two simultaneous waves, one going through the top slit, one going through the bottom slit, and these two waves (belonging to the same photon) interfere with each other. In other words, the single photon splits itself into two halves, with each half going through one slit. The half of the photon going through the top slit interferes with its other half going through the bottom slit. If three slits are used instead of two the photon will split itself into three interfering waves, if four slits four waves, five slits five waves, etc.

Ridiculous as it sounds, this is the position held by many quantum physicists, and the double-slit experiment with single photons is now a classic example of the so-called "quantum weirdness".

Instead of going into detail, I will claim simply that the propagation of light is a distributive process[5]. By applying the Time-Indifference Principle, the great "one-photon mystery" is immediately transformed into an expected result. In other words, it is not a mystery at all!!!

Such is the power of the Time-Indifference Principle!

The Time-Indifference Principle
as another evidence against probability

In hindsight, the "single photon mystery" is a mystery only because probability was assumed to be the working principle behind the phenomenon. By the principle of probability, each event is independent of all other events. This would make the formation of the long-term pattern very mysterious indeed.

But once the principle of symmetry is applied, the phenomenon is not only understandable, but in fact trivial. Thus, the success of the Time-Indifference Principle is another evidence against probability and in favor of symmetry.

A final note on space and time

Although coin tossing is a physical process, the attributes "heads" and "tails" have nothing to do with space and time. The same is true for the tossing of dice. For reference purposes, I will call these "non-physical distributive processes".

If we investigate the behavior of photons as in the double slit experiment, we will encounter a different kind of distributive processes because here distributive entities such as position, time, momentum, energy must be defined in terms of space and time. I will call these "physical distributive processes".

These two kinds of processes can be differentiated as follows:

Non-physical distributive process: The distributive nature of the process is caused by external manipulations (e.g., coin tossing action).

Physical distributive process: The distributive nature of the process is caused by the inherent correlation of space and time (manifested as position and momentum of the object, for example.)

Physical distributive processes are the dominant processes in quantum physics. Since space and time play important roles in physical distributive processes, the idea of "communication speed" (which I spoke against earlier) adds to the confusion, creating many apparent paradoxes that are still not satisfactorily solved at the time of this writing. I will address these problems in the book "The symmetry foundation of quantum physics". For now it suffices to say that with a proper understanding of space and time, all of the paradoxes of quantum physics can be easily solved./

NOTES:

1. One practical strategy to increase the randomness of computer algorithms is to use an external random source for inputs. For example, if the readings of a digital thermometer are accurate to, say, two decimal places and a particular reading is 57.231111, then 1111 can be used as the random input to the computer.

The alert reader will note, however, that the digital thermometer is itself a man-made machine. For this reason, its "random" part can only possess finite randomness.

Thus, the overall randomness is increased only because it is the combined randomness of two systems (the computer and the digital thermometer). Now by simply recognizing that the two systems are together a combined system, we can see that the overall randomness capacity still must be finite.

2. It is important to differentiate "synchronicity" from "synchrony". The later refers to the development of order from chaos by innate and live systems. For a good account of synchrony read "Sync – the Emerging Science of Spontaneous Order" (book), Steven Strogatz, 2003, THEIA (Hyperion books). It is interesting to note that, in this book, Strogatz seemed to doubt if synchrocity is valid at all.

3. "Synchronicity: An acausal connecting principle" (book), C.G. Jung, English translation by R.F.C. Hull, third edition, Princeton, 1973.

4. Imagine this. You ask a three years old toss a coin many times. Heads and tails show up randomly. The outcomes seem so patternless that there is no way you can remember the past outcomes or predict the future outcomes. Yet after gathering the outcomes of many tosses, you find all the ratio for heads is 0.5. This is very amazing (at least in our everyday's standard) because the seemingly unrelated tosses are somehow correlated across the barrier of time!

5. Strictly speaking, the claim that light propagation is a distributive process needs to be substantiated. I will do this in my next book "The Symmetry Foundation of Quantum Physics".

Written January 2003

Chapter 8

The Central Limit Theorem and the future of science

We will discuss the connection between individual events and "the system" that leads to the interdependency of single events. We will show why classical physics is the wrong foundation for the humanity sciences. We will show that the Central Limit Theorem may be the governing rule of the process of life itself.

We will show that the differentiation between deterministic and distributive phenomena is observer-dependent, and, in a sense, every phenomenon has both characteristics. We will show that deterministic chaotic processes are not distributive, but all quantum processes are. We will point out that the class of quantum processes includes many "ordinary" processes, including most of the processes in our daily lives.

We will show that the Central Limit Theorem averaging process is the reason for the abundance of normal distributions in our world. We will have a discussion on the role of individuals in a world that seems to be dominated by the Central Limit Theorem.

In closing we will make the statement that the Central Limit Theorem holds the key to the future of science and specifically in the unification of science. For this reason, we will make the prediction that it will become the most important theorem of science.

A. INDIVIDUAL VS SYSTEM

The partial inequality of individual events

In chapter 5, we deduced that individual events in a distributive process are inter-dependent. We will now determine what "inter-dependency" means and where it comes from.

Let's pretend that we are tossing 16 unbiased coins (N=16, p=0.5) in a process with k=3 where Heads is the chosen side.

Recall from chapter 5 that the maximum deviation is determined by:
$$|N_S\text{-}Np|_{max} = k\{Np(1\text{-}p)\}^{1/2} \qquad (1)$$
This gives us the result:
$$|N_S\text{-}Np|_{max} = 3\{16(0.5)(0.5)\}^{1/2} = 3(2) = 6 \qquad (2)$$
The average for Heads can be calculated with the formula for the mean value of a binomial process:
$$\mu = Np = 16 \times 0.5 = 8 \qquad (3)$$
From (2) we know that N_S can only be within the range of:
$$8 +/- 6 = 2 \text{ to } 14 \qquad (4)$$

This means no matter how many times these 16 coins are flipped, if we count the number of heads, we will find it to be always between 2 and 14. We will never have any of the following cases: 0 heads 16 tails, 1 heads 15 tails, 15 heads 1 tails, 16 heads 0 tails.

Now pretend that we have an extremely precise timer that times exactly when each coin settles into heads or tails. Since it is impossible for any two coins to settle at the same moment, we will know their exact sequence.

Let's look at the case where the first 13 coins have turned up heads. What can the 14th coin be? Answer: It could turn up heads or tails, because 13th heads is still 1 below the limit.

Now things get more interesting. If the 14th coins turned up tails, what would the 15th coin be? Answer: It could turn up heads or tails, because there are only 13th heads, still one more is allowed. But what if the 14th coin already turned up heads? Answer: The 15th coin has to be tails!!! More, the 16th coin will also have to be tails!!!

What we have just found is that in distributive processes, the events that come late may be at the mercy of the events preceding them. Depending on what already happened, the late comers sometimes have a choice, sometimes they do not!!!

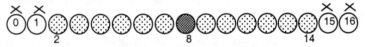

Figure 1: In this coin flipping experiment with the k-factor = 3, the minimum number of heads is 2, the maximum 14. We know for sure that the four combinations 0 heads 16 tails, 1 heads 15 tails, 15 heads 1 tails, 16 heads 0 tails can never occur. This is different from the Probability theory, which says that the k-factor is infinity and all combinations are possible.

This example sounds so wild that some of us may think that it is only a mathematical trick and does not reflect real life at all. The next example, therefore, comes from real life. Imagine ourselves observing a gift exchange party. There are 25 people and 25 gifts. A random draw gives each person a number that matches one of the gifts. It may be impossible to guess which gift out of the 25 will be taken by the first person. But the situation changes with the second person, because he or she cannot have the gift that the first person has taken. Likewise the third person cannot have the two gifts that the first two persons have taken, etc. Finally, after 24 persons have taken their gifts, we know with absolute certainty which gift the last person will get, because it is the only one left on the table. Thus, the degree of uncertainty is maximum with the first person, then reduces gradually and becomes zero with the last person.

The Central Limit Theorem and the future of science

We still have one interesting example left before making our point. Table 1 lists the maximum consecutive occurrences allowed as a function of the propensity p and the k-factor. In case some of us have forgotten, we have built two similar tables in chapter 5, the only difference is that we are using very small k-factor this time. Note that only the numbers in **bold** are feasible combinations. The reader may want to convince him or herself that p=0.4, k=0.01 cannot occur at all, although the maximum consecutive occurrence is shown to be 1. (Hint: This curious situation arises from the requirement that all numbers have to be rounded-down integers.)

TABLE 1: Maximum consecutive occurrences for small k-factor

k-factor	0.01	0.50	0.90	1.20	1.70	2.10	2.80
propensity							
0.01	0	0	0	0	0	0	**1**
0.05	0	0	0	0	0	**1**	**1**
0.10	0	0	0	0	**1**	**1**	**2**
0.20	0	0	0	**1**	**1**	**1**	**3**
0.30	0	0	**1**	**1**	**1**	**2**	**4**
0.40	1	**1**	**1**	**1**	**2**	**3**	**5**
0.50	**1**	**1**	**1**	**2**	**3**	**4**	**6**
0.60	**1**	**2**	**2**	**3**	**4**	**6**	**9**
0.70	**2**	**2**	**3**	**4**	**6**	**8**	**13**
0.80	**4**	**4**	**5**	**7**	**10**	**13**	**21**
0.90	**8**	**9**	**12**	**15**	**22**	**29**	**45**
0.95	**17**	**20**	**25**	**31**	46	**60**	**94**
0.99	**91**	**103**	**131**	**163**	**235**	**310**	**481**

Let's focus our attention to the case p=0.5, k=0.01 (which possibly represents the coin flipping action of a very well controlled robot.) The maximum occurrence is 1. If the first flip is heads, then we know immediately that all odd flips will be heads, all even flips will be tails. The first flip, then, contains the only uncertainty of the process; while the rest is completely deterministic.

After examining the three examples, we find:

-The 16-coin process is the most random of the three, because only in very selective cases that the last two coins are at the mercy of the previous 14.

-The 25-gift process is medium in randomness, because we can slowly eliminate the possibilities, and predict the exact outcome of the last event.

-The coin-flipping robot example is the least random because once we know the outcome of the first flip we can predict all other outcomes.

Thus, it is clear that randomness level is the key word in the inter-dependency of individual events. Since randomness level is decided by the k-factor, which is a system constant, the ultimate cause of inter-dependency of individual events is the system in operation.

The Probability theory has no consideration for the system because it assumes that the k-factor is infinity, making all systems equally random at the maximum level possible. This wrong assumption led to the wrong conclusion that all events are equivalent in a distributive process.

The reality of finite k-factor makes it impossible for all events to be treated equally by the system. Thus, the inter-dependency of individual events is a form of inequality imposed on the events by the system.

Therefore, instead of saying that individual events are inter-dependent, we will choose to be more precise and state that some events in distributive processes are not equal to the others. In today's politically correct language, the statement would be "Some events are more equal than others." This is the law of "partial inequality" that applies to individual events in all distributive processes.

A qualification needs to be made at this point. The picture we have just presented is the one seen by the Distribution theory. It is a complete picture of reality, according to the existing definition of science. Shockingly, it still turns out to be an incomplete description of Nature. We will return to this point in the section "the need for a new approach to science".

Systems and the Central Limit Theorem

As far as the system is concerned, it does not pay special attention to any particular event. All it cares about is that the rules of the system will somehow be enforced.

We can compare the system to an organization, single events to individuals in the organization. An organization must have rules to keep things under control. In order for a real life organization, say a country, to keep things under control it must evaluate the total situation and act accordingly. But what criterion or criteria does it use to evaluate the total situation? The average, of course! This is why we keep hearing things like "earning per capita", "average consumer spending", "average crime rate per 10,000 citizens", etc.

The rules of a real life system, then, are the Laws of Average. The word "average" is the very key word of the Central Limit Theorem. We already know that the Central Limit Theorem is the dominant force in distributive processes. We now see that it could be the dominant force of life itself!

System determinism and individual freedom

We have found that at finite randomness, some events are deterministic in the very strict sense of classical physics. It is not difficult to see that when randomness is reduced to zero, all events are classically deterministic and distributive processes cannot exist at all!

Thus, the determinism of classical physics must rely on the assumption that the universe is completely devoid of distributive process. The very existence of distributive processes, then, assures us that the deterministic picture assumed by classical physics is an incomplete description of Nature. In the correct description, there is always some level of randomness.

"Randomness" is a word with bad connotation. As far as individuals are concerned, we feel better with the word "freedom". Now that we know for sure that classical determinism is only a partial picture of Nature, we can be certain that at least some of us, hopefully most of us, are free from the stranglehold of "the system" and operate with a different set of principles. It follows that our current reliance on deterministic sciences to solve the problems of life is the wrong approach; which explains why the incredible progresses in science have not been translated into any measurable progress in our spiritual well being. If anything, science and technology have pushed us away from one another, making us more and more lonely and isolated.

The Distribution theory is equally guilty because it is a study of the system, not of the individuals. However, it opens the door of opportunity, because it tells us that a completely new scientific approach is needed to investigate the "force of one" of the individuals.

The need for a new approach to science

It is important to emphasize what we mean by "a new scientific approach". By existing definition, science is the objective observer of the universe. The problem is, an objective observer has to rely on a uniform standard while he gathers information. For this reason, science has no choice but to treat all observed objects as if they are equally passive. The all-important "cause and effect" connection, for example, can only be decided by the time factor. If B always follows A science would make the statement that "A causes B". This approach has been very effective, but it is expected to fail miserably in distributive processes. Why? Because distributive processes have no respect for time, as evident by the "time-indifference principle".

The Distribution theory is still an objective observer's view of "the system". This view tells us that the system affects the individuals and not the other way around. But is this a complete statement? Not necessarily. While there is no doubt that a computer program affects the numbers that it will output, we should keep in mind that the program itself is kept under control by the opposing forces of randomness and the Central Limit Theorem. Thus, in their own passive way, the outputs themselves do have an effect on the program.

The process of life is even more dynamic. Earlier we have used a country as an example of "the system". By a country, we mean its government. But what does a government consist of? People, of course. Thus, in life "cause and effect" does not go just one way. A government will affect the life of the citizens, but when a determined citizen with revolutionary ideas is voted in as the new president of the country, we can bet that the government itself will undergo revolutionary changes. Even the meaning of "determinism" becomes vague in this "two way" environment. It could mean a passive destiny (i.e., has to pay a new tax), but at the same time it could mean an unshakable determination fostered by previous events (i.e., determined to become a leader after seeing too much suffering by the people.)

The Central Limit Theorem averaging process has been presented by the Distribution theory as if it is an invisible power that spares no individual event. But could it be that, in life at least, the Central Limit Theorem is at times the rule exercised by the people for or against the system? Were the massive anti-government uprisings in Eastern Europe in 1989 an averaging act of "the system", or that the people decided to average things out themselves?

It can be seen that in the investigation of individual events in distributive processes, the old method of "objective observer" becomes very fuzzy, which defeats its own purpose. But can we really replace the "objective observer" model and still have science? What will the new approach be? The writer is looking forward to the opportunity to offer his answers to these challenging and exciting questions in the near future.

B. QUANTUM PROCESSES

Random and deterministic scales

Some classical physicists may argue that most processes are deterministic. We will proceed to prove that this is not the case.

To a bacterium, a variation of order 0.1mm is huge. To an ant, it is significant. To us human beings, it is insignificant (unless when we study science). To a giant dinosaur, it is the same as no variation. Thus, by going up in scales at some point randomness will change to determinism. Conversely, by going down in scale at some point determinism will change to randomness.

We conclude that determinism and randomness are simultaneous realities in all processes, and the difference arises from the scale chosen by or imposed on the observer. This point is extremely important, as a detail which is considered negligibly small from a system standpoint (or the viewpoint of another person) may change the life of a particular person forever.

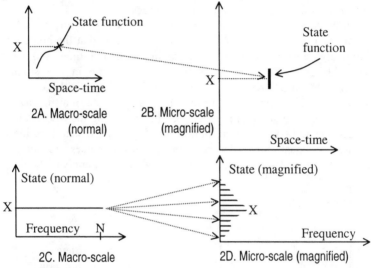

Figure 2: When details are important we find out that even the so-called deterministic processes are distributive in nature. This can be seen by amplifying a seemingly deterministic process and look at the details. Classical physics, then, only valid when the details can be overlooked. This is acceptable in many cases with innate objects, but the same cannot be said about life. That is because a detail that is considered small from a system standpoint may change an individual forever.

Following tradition, we will describe a process by its "state function". In figure 2A, which is the macro-scale description of a process, for each point in space-time there is only one point in the state function. However, if we keep magnifying the vertical dimension, eventually we will get to the micro-scale of figure 2B. At the micro-scale, the state function is no longer single-valued. Instead, for each value of space-time, the state function could take on any value within a relatively wide range (bold vertical segment in figure 2B). Thus, one point X in the macro-scale has become a line segment in the micro-scale, and the process has changed from being deterministic to being indeterminate.

In the macro-scale, if we repeat the same space-time condition many times and plot state vs. frequency we will get a horizontal line segment of value X and frequency N (see figure 2C). However, in the micro-scale, we will find that this same set of data do not fall on the same point X. Instead, it forms a distribution centered at X, which may or may not have the shape of a normal curve (see figure 2D).

Differentiating chaotic and quantum processes

Why does a single point X in the macro-scale expand to a line segment in the micro-scale? One possible reason is that the process is extremely sensitive to small changes in space and time, and what believed to be the same space-time conditions are in fact slightly different, leading to results different from X. Processes in this category are now called "deterministic chaotic processes". Deterministic chaotic processes can be successfully analyzed by increasing the resolution of space-time and taking into account of the fact that the state function is highly non-linear. Since deterministic chaotic processes are not true distributive processes, our subsequent analyses do not apply to them.

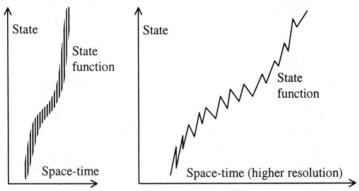

Figure 3: Chaotic process becomes deterministic when we increase the resolution of space-time and account for non-linearity. It is, however, not true that all distributive processes can be resolved into a deterministic chaotic process. In fact, if we keep going down in scale, at some point deterministic chaotic processes will also become distributive. The scale at which a process is distributive will be called the quantum scale. It is more correct to say that each process has its own quantum scale. If our interests are in the quantum scale, we are dealing with a quantum process.

At the time of this writing, many chaotic theorists believe that all distributive processes are deterministic chaotic processes (and therefore use the shorter term "chaotic processes" to refer to all of them.) We will argue that this is not the case.

Note that the state function of a deterministic chaotic process is continuous. The state function for a coin flipping experiment, on the other hand, is discontinuous (with only heads or tails and nothing in between). No matter how well we measure the mechanical parameters involved in a coin flip, the measurement error can never be reduced to zero. Since the measurement error is finite, there will always be cases where the deciding factor is immeasurable, and it is impossible to tell whether the coin will digitize to heads or tails. Thus, coin flipping is not a deterministic chaotic process.

We will call distributive processes with discontinuous state function "quantum processes" for reasons that will be clear when we discuss quantum mechanics.

If we keep going down in scale, at some point we will find deterministic chaotic processes also become distributive. We will call the scale where processes become distributive the quantum scale. It is clear that each process has its own quantum scale. If our interests are in the quantum scale, we are dealing with a quantum process.

Discontinuity as criterion for quantum processes

The word "quantum process" may scare a few readers, it is therefore important to emphasize that "quantum" simply means "discontinuous". While the word "quantum processes" usually refers to the microscopic realm of the atoms, we do not have to go down that far, because quantum processes are present everywhere in our daily life.

Coin flipping is probably the most familiar quantum process of all. It is quantum because the only two choices available are heads or tails, not something in between. By increasing the number of coins, we have more choices, and in essence make the "quantum gap" smaller. This can be seen by "normalizing" the two extremes into -1 and $+1$.

One coin: Heads = $+1$, tails = -1, quantum gap = $+1-(-1)=2$

Two coins: HH=$+1$, HT=0, TT=-1, quantum gap = $1-0=0-(-1)=1$

Three coins: HHH=1, HHT=$1/3$, HTT=$-1/3$, TTT=-1, quantum gap $=1-1/3=1/3-(-1/3)=-1/3-(-1)=2/3$.

Etc.

We know, however, that there is a limit in this gap reduction process because a completely random system simply does not and cannot exist in real life. Thus, a quantum gap will always be non-zero, and coin flipping will always be a quantum process.

But coin flipping is admittedly a boring process, the curious reader would ask "What about real life? My life?" The surprising answer is that life in general and individual lives in particular are very much quantum. In fact, it is not an outrageous overstatement to say that life is all quantum!!!

To illustrate this point let's follow the life of a female person from the beginning of her life. She was born female, the result of a quantum choice by Nature (or by her parents in this new age where sex could conceivably be manipulated.) In her childhood, she was taken care of by her mother and hardly knew her father because of a complex quantum choice evolved in the general culture where she was born. When a child, she preferred red shoes and white dresses, and felt very strongly about these quantum choices. Her father made a quantum choice between three schools and ending up picking the one closest to home for her. She got mostly A's and B's in the quantum system of grading imposed by the educational system. As a teenager she liked two boys, but decided to make a quantum choice to go steady with one. After graduation from college, a new man entered her life by accident. She had to make a quantum choice between her long time boyfriend and the new but charming person. She finally chose one and became his wife. After marriage, she had to decide between two quantum choices: Pursuing a career or staying home and dedicate her effort to home making. After deciding to pursue a career, she had an excellent job offer, which would require her to relocate. Her husband, being less capable, had to go along with her but did not feel comfortable with the fact that his life had been dominated by his wife. After striking a conversation with a timid girl in a social gathering and sensed her come-on gestures, he had to make a quantum choice between staying faithful to his wife and committing adultery. While he was still undecided, his wife heard the rumor that her husband had an affair and had to make a quantum choice between trusting her husband or making a big deal out of it at home, etc.

We can go on and on, but the point should be extremely clear by now: Quantum processes dominate our life, whether or not we choose to pay attention to them (which is, again, another quantum choice.)

The role of resolution limit in quantum processes

As it turns out, quantum processes are all about resolutions, or more correctly, the inability to achieve them. The word "resolution" here was used in its physical meaning. We hear it often in the industrial world as "tolerances", "precision", etc. Thus, as the reader may have guessed but probably found it hard to believe, most industrial processes have quantum processes imbedded in them.

Take the case of a construction contractor. When he orders a batch of metal rods, he may simply specify them as being 10 meters long. There seems to be nothing quantum here, but would 9.5 meters be acceptable? Probably not! Thus, although not specified, there is an acceptance range of say between 9.9 and 10.1 meters (the later comes from the manufacturer who is in business for profit not charity, since the contractor may not mind at all to receive 20 meters rods, having to pay only for 10 meters.) The manufacturer's specification becomes 10.0+/-0.1 meters, giving an allowable range of 0.2 meters. As a rule, manufacturers never want to spend more than they have to for any tools, so it is most likely that the "precision" of the cutting tool will be smaller than the tolerance range but only very slightly, say 0.18 meters.

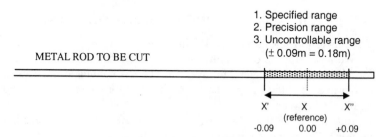

Figure 4: In this case, simply because we are interested in the distribution of cutting locations of many metal rods, the very low-tech metal rod cutting becomes a quantum process. This shows again that the differentiation of deterministic and quantum processes is observer-dependent.

The working meaning of this precision is that if we take a point X as the reference point, the actual point where the rod is cut will be somewhere between X' (at 0.09m from the left of X) and X" (at 0.09m from the right of X).

A cut at X would give a perfect rod of exactly 10.00 meters long. The software programmer will therefore set X as the desired cutting point, with the understanding that the actual individual results will most certainly deviate from X, and the only thing that he can tell about the next cut is that it will most likely be confined in the +/-0.09m range between X' and X".

The fact that the programmer cannot predict where the very next cut will be rules out the possibility of a deterministic chaotic process, because reducing the resolution is not an available option (due to budget constraints). Thus, this very crude process of metal rod cutting turn out to belong to the "elite" class of quantum processes, long thought to be exclusive to those exotic elements such as photons and atomic particles.

Note that 0.09m is 90mm, not a small number considering that 1 nanometer (0.000001mm) is the accepted unit measure in the high technology industry at the time of this writing. Yet (low-tech) metal cutting to within ± 90mm is quantum, and (high-tech) overlay to within ±0.01mm is not. This confirms our earlier observation, that deterministic and distributive processes are matters of relative scales.

As a general rule, whenever the range of interest is smaller than the resolution limit (i.e., the smallest resolution possible), we have a quantum process, regardless how big or small the typical measure is. This is understandable, because when a process exceeds its resolution limit it will appear as discontinuous, which qualifies it as a quantum process.

C. THE AMAZING CENTRAL LIMIT THEOREM

The power of the Central Limit Theorem

We now return to our familiar subject of coin flipping. Let +1 stand for heads, -1 for tails, and 0 for average. There is nothing average about a coin toss. It will turn up either heads or tails, for either +1 or −1 and nothing in between.

But what if we flip 100 coins at a time? Now suddenly average has meaning. For example a toss giving 55 heads and 45 tails gives a net of +55-45=+10. This we can call the "unnormalized" sample value. Unnormalized values are inconvenient because if the total number of coins are not known we have no idea what the numbers mean (a flip of 1000 coins with 505 heads and 495 tails also gives the same unnormalized value of +10). It is therefore common practice to divide the net result by the total number of coins, giving the "normalized" value, which in our daily language is known as "average". Thus the corresponding average for 55 heads and 45 tails is (55-45)/100=+0.1; for 505 heads and 495 tails it is (505-495)/1000=+0.01. Note, however, that unless the total number of coins is infinity, which is impossible to achieve, the average can only take on a finite number of values. For 100 coins there are 201 possibilities ranging from −1 to +1, with increments of 1/100=0.01. For 1000 coins there are 2001 possibilities ranging from −1 to +1, with increment of 1/1000=0.001.

It should be well known to the reader by now that if we take each average of 100 coins as a single data point, compile many of the same, and plot them; we will get a curve that is approximately normal. We know this is the case, thanks to the Central Limit Theorem.

What about the metal rod cutting machine that we have used as an example of quantum processes earlier? Even the simplest metal cutter has at least two moving parts (the driver for the blade, and the blade itself). In a perfect world each cutting action is a perfect repeat of the one before it. The problem is: The world is not perfect. The random location of the next cut, then, is decided by a complex combination of many random factors, of which some may be very erratic some may be much less so. Thus, it is the combined state of a multitude of random factors that decides the random location of the next cut. But what is a "combined state"? The answer is: The unnormalized value of all relevant random factors!!! Thus, if we compile many cutting locations and plot them we will get a distribution of the actions of the cutter.

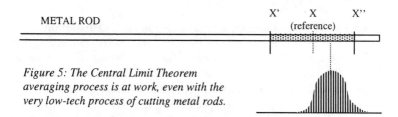

Figure 5: The Central Limit Theorem averaging process is at work, even with the very low-tech process of cutting metal rods.

Will this distribution be irregular? No, it will be approximately normal, though we cannot guarantee that its peak will coincide with the ideal cutting point as the programmer has hoped (this is called the "mean-shift" problem in manufacturing.) Why do we know that the distribution will be approximately normal? Again, because the Central Limit Theorem has told us that, regardless how irregular or unknown the factors are, as long as there is no systematic change, their average action would form a normal distribution. This is truly amazing, considering that it is impossible to tell where the cutting blade will be when it performs the next cut.

This may be too much of a good thing, but the writer feels the urge to inform the reader that we human beings are no exception to the Central Limit Theorem. There is no "average" human being, but whenever we perform any task, our action is the instantaneous average of the many, many factors that are relevant to the task, even when we are not aware of this fact at all. Catherine decision to buy her expensive dress, for example, seems to have come at a moment of impulse; but it is still an instantaneous combined average of the many factors A, B, C, D, E, F, etc. that operate inside her heart, her mind, and her subconscious. Other women that Catherine will never know but are in the same culture with her also have the same set of factors operating inside their heart, their mind, and their subconscious. Just that their instantaneous average actions may be different from Catherine's. As long as known systematic differences are properly accounted for, the shopping pattern of the women in Catherine's culture will form a normal distribution, as predicted by the Central Limit Theorem.

Or take the case of a standardized national exam. The performance of each student is an instantaneous average of the many factors that are relevant to their exam taking ability: Their upbringing, parental guidance at home, the educational system of the country, personal dedication, intelligence level, psychological stability under stress, memory power, cramming skills, etc. It is no secret that, after discounting systematic differences, these test scores form normal distributions. The reason is –again- the averaging process dictated by the Central Limit Theorem.

The Central Limit Theorem as reason for normal distributions

The few examples cited above may give the readers the feeling that they were carefully selected to exaggerate the power of the Central Limit Theorem. The fact is: A majority if not most distributive processes in the world are known to obey the normal distribution.

There have been numerous speculations that this curious fact is the work of the Central Limit Theorem. The logical problem with these speculations is that the probabilistic version of the Central Limit Theorem is only meaningful with very large sample sizes, but the number of factors involved in many naturally occurring normal distributions is relatively small. The Distribution theory overcomes this difficulty with symmetry, making it possible for the Central Limit Theorem to work with relatively small sample sizes. Thus, thanks to the Distribution theory we can conclude with certainty that the Central Limit Theorem is indeed the rule behind the numerous normal distributions that we observe in Nature.

Take people's height, for example. If we tally the heights of a large group of adult males in a relatively homogeneous population, we can predict almost with certainty that the distribution will form the familiar bell-shaped curve of a normal distribution. We also see similar normal behavior with a large group of adult females from the same population. If we combine data for the two groups in the same graph, however, we can be sure that the distribution will be bimodal (i.e., having two maxima).

Height is a result of many factors, among them and probably most significant of all is genetic. Recently science discovered that the nature of genetic is binary, but this is not the point here, as other factors that affect height may not be binary. If we can somehow quantify these factors as contributors to height and tally them separately, the resulting distributions for most of them are expected to be irregular.

Note, however, that each height is no more than one trial (with large sample size) by Mother Nature via the growing process of human bodies. We can safely assume that most relevant factors are randomly chosen in this growth process because, after all, Mother Nature should follows Her own rule: The equal opportunity principle. Thus, by the

Central Limit theorem, height should be normally distributed.

(By the way, the version of the Central Limit Theorem used in this example is credited to Liapounov. This version states that if we take many samples, each from a different distribution, average them and plot them we will get a normal curve when the number of samples approaches infinity. The Distribution theory allows us to relax the later condition to "large enough sample size". The reader may want to prove Liapounov's version of the Central Limit Theorem as an intellectual exercise.)

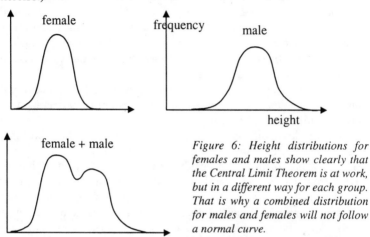

Figure 6: Height distributions for females and males show clearly that the Central Limit Theorem is at work, but in a different way for each group. That is why a combined distribution for males and females will not follow a normal curve.

The reason for the difference between the height distributions for male and female can be easily understood: There is a clear systematic genetic difference between the two groups. It is a known fact of mathematics that the averaging process cannot equalize systematic factors.

Since we have represented a "low-tech" example with the metal rod cutting process, it is only fair to balance the situation with a high tech example of the Central Limit Theorem. "Steppers" are very high-end cameras that cost millions of US dollars each. They are used to define many "layers" on top of one another, one at a time, in a process called "overlay". In the overlay process, it is important to keep the alignment error between two successive layers within certain specifications. These specifications are usually defined in terms of 3 sigma's, meaning that if the data are fitted into a normal distribution, the 3-sigma number has to be below the specification. Here normal distribution is not just assumed, but taken for granted as a solid fact. That is because it is known to all engineers in the trade that the distributions of overlay data are indeed normal.

A careful analysis again points toward an averaging process dictated by the Central Limit theorem. Every component of a stepper has its own ideal position or/and configuration which, as a rule, it always deviates from. The individual deviations combine in a complex way to give the stepper its overall performance. Each overlay data point, then, is an instantaneous average of all of the deviations. By the Central Limit theorem, the distribution for multiple overlay data points is normally distributed.

By applying the same analysis to any distributive entity that comes to our mind, we will find that the normal distribution is not the exception, but the rule of the universe! As long as the final measure is a complex combination of many variations, its distribution should be normally distributed. We dare to make this statement because the Central Limit theorem demands it.

Thus, Central Limit Theorem averaging is the hidden process of the universe, and the regular occurrence of the normal distribution is a clear manifestation of this rule.

We will leave this section with a qualification, that there exist many non-normal distributions. One example is the distribution of particle sizes in a pile of dust. Smaller dust particles are formed by breaking away from larger ones, so dust particles are not equivalent in their mechanism of formation. However, while the Central Limit Theorem averaging process cannot be applied directly to particle sizes, it still works after the systematic factor (of smaller particles breaking away from bigger ones) is taken into account mathematically. The result is a distributive property called "log normal". It is interesting to note that "log normal" applies not only to dust (which no one cares about) but also to the prices of stock shares (which everyone cares about).

The meaning of individual events and individual decisions

In his effort to present a strong case for the Central Limit Theorem, the writer may have caused grave disappointment to some readers. After all, no human being wants to feel as if he or she is just a face in the crowd, or worse, a little dot under the bell-shaped normal curve. We all want to feel that we are special individuals, not necessarily great, but distinct and proud. In short, we don't want to be CLT'ed to nothingness!!!

To address this concern we will restate the point that we made earlier, that individuals operate on a set of rules independent of the Central Limit Theorem and may even cause the Central Limit Theorem averaging process at times. (This statement goes with an "almost" qualification, as we will see later.)

The Central Limit Theorem and the future of science

The big mistake of the Probability theory was that it assigned the long-term averages of distributions to the probabilities of individual events. There were historical reasons for this mistake, but we will not dwell into the past. It suffices to say that the concept of individual event probabilities has been proven irrelevant in the study of distributions. This does not mean that individual events are dictated by collective distributions, it simply means that any conclusion regarding an individual event is outside the scope of the Probability theory.

To see why this is the case, let's consider the possibility of getting N consecutive unbiased coins turning up heads. According to the Probability theory, the probability for this to happen is $p=1/2^N$. If N is very large p will be very small; but can we say that it is impossible to get N consecutive heads? No, the only conclusion that can be made with the Probability theory is that "it is very unlikely to get N consecutive heads." This leaves the possibility of N consecutive heads showing up at some unknown point in time, and that could be the next time we try to flip N coins.

By generalizing to the case N approaching infinity, the only conclusion that the Probability theory can make about an individual event is "Every possibility is qualified as the next reality!" Thus, we can emphasize once more that long-term averages (taken as probabilities) are meaningless predictions for single events.

This is where the Distribution theory again proves itself superior to the Probability theory, because we can in fact make certain deductions regarding individual events with the Distribution theory.

Let's return to a coin flip. We will ask the seemingly stupid question: "Is the flip completely random?" The Probability theory would say yes, but we now know that this cannot be the case; because if it were, the averaging process of the Central Limit Theorem would not converge.

Thus, the fact that the Central Limit Theorem works is a proof that our decisions are not completely random. But do we want our decisions to be completely random? The writer thinks not, because complete randomness is the same as non-personality. Thus, the Central Limit Theorem does not deny personality. Quite the contrary, it confirms that personality necessarily exists.

Maybe we should instead go the other way and ask how deterministic our decisions are. According to the Distribution theory the number of sigma's is fixed and depends on the "randomness level" and "randomness capacity" of the system in operation. Thus, all individual events are confined within the range of k sigma's. This is not total determinism, but it is still a form of determinism, because we know that no individual events can deviate from k sigma's.

SYMMETRY AND THE END OF PROBABILITY

For each individual event, there is a corresponding set of specific information that is unique to it. With this set of information, we may be able to reduce the number of random factors and get a new distribution, much narrower than the original one. This is possible according to the rule of the Central Limit Theorem, because when the number of random factors is reduced, the resulting normal distribution becomes narrower. Obviously, there has to be a limit to this reduction process and there will always be some degree of uncertainty at the end. This final degree of uncertainty again depends on the randomness level and randomness capacity of the system that governs the individual event in question. This means there may be events that are, for all practical purposes, completely deterministic; while others possess various degrees of determinism (or various degrees of uncertainty, depending on the observing angle of the beholder.)

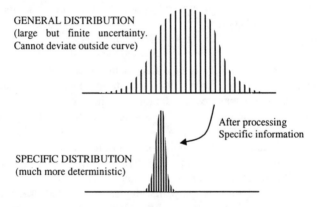

GENERAL DISTRIBUTION
(large but finite uncertainty.
Cannot deviate outside curve)

After processing
Specific information

SPECIFIC DISTRIBUTION
(much more deterministic)

Figure 7: By processing individual information, we can reduce the degree of uncertainty and narrow the deviation range, thus increasing the deterministic level of individual events. However, in most cases this reduction process will have a limit and there is always some residual uncertainty. To account for this uncertainty would require a new approach, a new science

Total determinism is probably depressing to most of us, but paradoxically partial determinism is a necessity. After all, there are times when we want to guess correctly what our loved ones will do next. It would be frustrating if their personality changes often and without notice. If we did not believe in partial determinism, there would be no justifications for the teaching of psychology, psychoanalysis, and sociology in schools.

The field of fortune telling, astrology in particular, puts a clear stress on fate and destiny, which is very close to total determinism.

Despite strong opposition by scientists and even religious leaders, fortune telling keeps on living and prospering. This fact tells us that a significant percentage of our fellow human beings believe in determinism. They have the right to do so, and under certain circumstances, they may even be right.

In summary, the Distribution theory tells us that life, our individualism included, must have certain degree of determinism. This implies, however, that certain degree of freedom must also exist. More detailed statements regarding the determinism and freedom of life would require a separate study, which the writer hopes to present to the readers in a future opportunity.

Since we have been discussing individual coin flipping events, before leaving this section the writer would like to present a scientific puzzle that has been around for thousands of years. In the modern I-Ching method of divination, 3 coins are tossed each time for a total of 6 times. The 6 resulting heads/tails combinations are used to construct a hexagram that supposedly has predictive power. Since this is not a book on I-Ching there will be no detail given on what the four possible combinations (HHH, HHT, HTT, TTT) may mean.

According to the Probability theory, coin flipping is a completely random process. This would make the I-Ching procedure of coin flipping sound stupid; but we now know total randomness cannot exist. Thus, there may be some acceptable logic behind the I-Ching procedure. The question: "What could the logic be?"

No reward is promised, but the intellectual satisfaction associated with being able to solve a challenging puzzle is possibly the greatest reward of all. It is necessary to emphasize, however, that this is not a trivial problem (scientific problem, that is.) The writer has been working on it for more than 5 years, and although being very close, he still has not solved 100% of it.

D. THE FUTURE OF SCIENCE

The Probability mistake with the Central Limit Theorem

Let's return once more to the issue of event probability. When the Probability theory states that the probability for the next coin toss to turn up heads (or tails) is 0.50 it should have said: "Assuming there is no process bias, the long run average for heads would be 50%." Likewise, the probability for one million heads in a row should be restated as: "Assuming there is no process bias, if a million coins are flipped each time, the long run average for all-heads combination would be $1/2^{1000000}$ of all one-million-coin combinations." This is still most likely a wrong statement (because of the limited "randomness capacity" of real systems), but at least it would spare the Probability theory the absurdities that deny it the status of a serious science.

The key word, then, is again "average". What does this key word remind us of? The Central Limit Theorem, of course.

Thus, the Probability theory has unintentionally mixed up between the CLT "average combination", which only has mathematical meaning in many cases, with individual combinations. Worse, it repeated the same mistake with all other distributive processes. By so doing, the theory created its own nightmare, because accepting probability is to give up the guarantee for convergence, as we have shown in chapter 2.

In hindsight, the damage caused by the Probability theory was two-fold; because with (misinterpreted) probability occupying the center stage and minimizing the credibility of the Probability theory, it was impossible for the scientific community to recognize and appreciate the somewhat similar but much, much more significant concept implied by the Central Limit Theorem. As a result, the Central Limit Theorem has been limited to the minor role of a beautiful mathematical theorem, admired for a few brief moments, applied on one or two rare occasions, then forgotten. Before working on this book, the writer had only encountered the Central Limit Theorem once, and that was in a statistics class.

As it turns out, the Central Limit Theorem is the theorem of the Distribution theory, as we have seen in the last several chapters:

1. It generalizes all distributive entities as binomial combinations, with predictable mean, sigma, and even convergence limit.
2. It is the governing rule of many natural phenomena, in industry, in technology, and in all aspects of human lives; giving them the normal distribution that we are so familiar with.

Perhaps the most damaging impact of the probability concept was in the development of quantum physics. Probability was introduced to quantum physics in 1926 by Max Born as a necessity. It became the backbone of this exciting branch of science. Quantum electrodynamics, the most successful quantum theory to date, is based completely on probability. The theory works beautifully when it comes to calculations; but the concept of probability bothered even Richard Feynman, who was credited with the most useful form of the theory.

As it stands today, quantum physics is still plagued with a mountain of unanswered questions. Most of them have something to do with the concept of probability.

In the next book we will see that the key in solving quantum puzzles and finally making this great branch of human knowledge a true science is again the Central Limit Theorem.

The role of the Central Limit Theorem in quantum physics

Readers who are familiar with quantum physics may immediately raise their objections to the claim that we have just made, and they have good reasons. After all, our discussions have been very much confined to the normal distribution, which led to the Central Limit Theorem. Since the probability distributions in quantum physics are in general not normal, why should the Central Limit Theorem have anything to do with them?

We note, however, that the formation of quantum distributions is itself a distributive process. Without the symmetry implied by the Central Limit Theorem, there would be no guarantee that quantum distributions could be reproduced. Thus, the very fact that quantum distribution are routinely reproduced confirms that the Central Limit Theorem has a major role to play in quantum physics.

In addition, we will find that the theory of quantum electrodynamics (QED), which still is considered as highly speculative at the time of this writing, can be accounted for logically by a modified version of the Central Limit Theorem. The Central Limit Theorem will finally justify the QED procedure, specifically Richard Feynman's controversial sum-over-histories procedure, as a part of standard science.

The conceptual problems with quantum distributions are much simpler than those with QED. They will be dealt with as they arise and will not be elaborated further here. It suffices to say, however, that all probability-related parts of quantum physics that are considered conceptually unsatisfactory at the present will become compatible with reality after being corrected by the Distribution theory in general and the Central Limit Theorem in particular.

The Law of Average and the future of science

The ultimate dream of science is the unification of all scientific principles under one roof. Some of us must have heard of "The Grand Unified Theory" of physics (GUT) which is in existence today. As it turns out, GUT is more a promising name than a reality. It is believed that there are four major forces in the universe: The force of gravity that we are all familiar with; the electro-magnetic force that governs the actions of electricity, magnets, and photons; the strong force that explains nuclear reactions; and the weak force that explains nuclear radiation. GUT claims to have unified the last three and is on its way to "bring gravity home", so to speak. A careful examination, however, reveals that the partial unification claimed by GUT still have many loose ends. In addition, even the most optimistic GUT proponent would agree that the theory might never have a satisfactory solution for gravity.

SYMMETRY AND THE END OF PROBABILITY

Superstring, the newest fashion of science at the time of this writing, believes it will beat GUT to the finish line in the unification marathon because gravity is built in with its first principles. At the same time superstring specialists complain that the existing mathematical techniques are "not ready" to solve the extremely complex equations that arise from the theory. This means it will take a long time before Superstring has any verifiable result of significance to share with the scientific community.

Let's assume for argument's sake that some day either GUT or Superstrings, or both, are successful in achieving their goal. Would their success complete the unification of physics? No! Because neither GUT nor Superstring has room for the very important class of so-called random phenomena, which require separate treatments by Chaotic theory, Catastrophe theory, etc.

We have not even mentioned the humanity sciences such as psychology, sociology, and economics where fuzzy logic is waiting for its heyday. It is impossible to see how these sciences would blend in with any theory in physics today.

Facing such enormous difficulties, many have argued that the unification of science would forever be an elusive dream. Specifically, they stress that it is theoretically impossible to ever bring natural sciences and humanity sciences together. The logic goes like this: Natural science is based on logic and mechanical principles, which are predictable; humanity science is based on emotions and reactions, which are unpredictable. The contrasts are simply too big to overcome.

As it turns out, the two branches of sciences do share a very important first principle. There is one old "law" that has been around for as long as anyone can remember. It is the celebrated "Law of Average". We heard it once in a while in daily conversation (and borrowed its name in an earlier chapter.) It is our consolation when things do not go our way or some bad person has done something wrong to us. We would say "I believe in the law of average. Things will get better," or would curse "The law of average will take care of him". We all have moments when we feel the law of average must be an important governing force in life, although we cannot prove it.

But what is the "Law of Average"? The reader can guess the answer from several chapters ago. It is the same Law of Average that we have been discussing many time in this book; and it is the working mechanism of The Central Limit Theorem. Thus, with the Central Limit Theorem science has in its hand a powerful tool that is equally applicable to both natural science and humanity science. It just didn't know how to use this powerful tool for three hundred years.

Although it is still too early, the writer will go ahead and make the prediction that the ultimate unification of science, and this of course means all of sciences, natural as well as humanity, will soon be possible. That is because the missing link has been re-discovered, and it is the celebrated Law of Average disguised under the name of the Central Limit Theorem.

The Central Limit Theorem will rise from obscurity, and it will do so in a big way. Not only that it will stand side by side with the great deterministic laws of Newton and Einstein, it should even surpass them. Because while Newton-Einstein laws and the Central Limit Theorem are equals in the world of innate matter and energy, only the Central Limit Theorem holds the key to the ultimate goal of science, which is a clear understanding of the heart and mind of the human race and its destiny in the universe./

Written September 2001
Revised November 2001, December 2002
© DangSon. All rights reserved

END OF BOOK 1

Please proceed to book 2
also written in layman language

"The Symmetry Foundation of Quantum Physics"

Gambler's
Wisdom beyond probability

TO LIVE IS TO GAMBLE

Although we may not realize it, we are all gamblers, just to different degrees, on different matters, and in different ways. If you're not convinced that this is the case, just consider the following examples:

Choosing between journalism and engineering for a college major, isn't this a gamble? Choosing between partying on a weekend or studying for an exam, isn't this a gamble? Choosing one among several people for a husband or wife, isn't this a (huge huge) gamble? Choosing between staying on a dead-end job or leaving for a relatively unknown future, isn't this a gamble?

The list goes on and on. Fact is, *to live is to gamble!* We may as well accept this fact so that we can deal with the unpredictability of life more efficiently.

So whether you're a high roller who frequents casinos to seek the thrill in blackjack, roulette, crap, poker, slot machines, and many other exciting man-made games of chance; or a conservative person who just wants to lead a quiet life, this section may have something useful for you. Because, in one important respect, all games are the same: They reward winners and punish losers. Once you're already in a game for whatever reason, you may as well try your best to win.

WHY AND HOW THE PROBABILITY THEORY IS WRONG

When we talk about games of chances, the first word that comes to mind is Probability!

If you have read the other parts of this book, I assume you will agree with me that we need to look beyond probability to handle the games of chance in a scientific manner. But if you haven't, I hope you will buy my argument after reading the following description of the three problems associated with the theory of probability.

Probability problem 1: It fails to account for process variations

To me, the most puzzling defect of the probability theory is its failure account for a factor that may have pronounced effect on the outcomes of distributive processes: The process itself!

In experiments with coins, for example, the probability theory assumes that the construction of the coin is the only deciding factor.

This assumption is hopelessly incomplete, as you will see in the following two examples:

1. How to change the probability of a coin with technology

I will pretend that I have a perfect US quarter in my hand. Now if someone suddenly asks me "what is the probability for the next toss to be heads", I possibly would answer "50/50!" That is because I, like everybody else, have been conditioned to accept the probability theory as a scientific truth for so long that it is hard to rub off the habit. Also because I have tossed enough coins in the past and confirmed that the long-term ratio for heads is about 0.5.

I will now change the condition of the experiment! Instead of tossing the coin myself I will feed it to the holding plate of a state-of-the-art robot arm of today and let the robot arm do the tossing for me. Will the long-term record of many similar tosses still give me 50% heads occurrence? The answer is "Not necessarily!" because, with the advanced high technology of today, it is possible to program the arm so that it will give me whatever long-term ratio that I desire, including heads 100% of the time and tails 100% of the time.

This very realizable "thought experiment" forces me to re-examine the concept of probability. When the probability theory was first formulated, there was no high technology and the most natural way to investigate long-term ratios was to toss a coin or many similar coins. It so happens that the act of coin tossing by human beings does not introduce any new factor into the long-term ratios of heads and tails. So it was luck that made the statement "the probability for heads is 0.5" appear correct in the old time; but with modern technology, luck has finally run out on this signature statement of the probability theory.

2. How to change the probability of a coin by spinning it

There is a much simpler way to manipulate the long-term ratio of a coin. You can do this yourself, by spinning a coin instead of tossing it.

I discovered this by accident in the year 1978, if I recall correctly. At the time I was a graduate student, but not the kind of graduate student that every advisor wants to have. Most of my time was either well spent for a good cause (e.g., community works) or wasted happily in bars and pool halls.

One late morning I was smoking and drinking coffee in the Valli Pub in Dinky Town, Minneapolis, Minnesota to recover from a hang over the night before. Having nothing better to do, I started spinning a quarter on the bar's table, waited until it tumbled around and settled, then picked it up and spinned it again. It took a while before I realized that I saw many more heads than tails. I ended up making a lot of spins, first randomly then a little more organized. I found I was able to skew the long term ratio to about 70/30 in favor of whatever side I chose. I told this to my foosball partner later in the day. Cut a long story short, he ended up winning many bar bets from our fellow college students, who were told by their probability professors that the odds were 50/50.

Finally in 1999, some twenty odd years after this minor episode I randomly picked up a book during a business trip in Singapore (I recall vaguely that the book title was "Randomness", but I forgot the author's name.) To my surprise the book mentioned coin spinning experiments; and it confirmed that long-term ratio in coin spinning experiments usually deviates from 50/50. I may add that reading this book brought back old memories and motivated me to re-examine the theory of probability even more seriously.

I will not tell you a "proper procedure" to spin a coin, because by doing so I would rob you the pleasure of figuring this out by yourself. However, I guarantee you that watching a spinning coin finally tumbling down and showing up the side you picked almost every time will be one of those most puzzling and satisfying moments that you ever experience. It makes you feel as if randomness is only a mask, with determinism hiding right behind it.

Probability problem 2:
It fails to predict the absence of long series

Because the computer is the brain behind many casino games, I will now use a result obtained on a computer as evidence against probability. As I reported earlier in this book, Mr. Hanspeter Bleuler, a long-time friend of mine, ran a long program in November 2002 to search for consecutive strings of unbiased 0's and 1's. Of the total of 141.3 billion 0's and 1's he found the longest string to be 28. This string occurred 521 times; in good agreement with the 526 predicted by the probability theory.

A probability expert would tell you that for the same total of 141.3 billion 0's and 1's you should expect roughly:
263 strings of 29
231 strings of 30
115 strings of 31
etc.

Adding all of these together, the probability theory predicts roughly 526 strings of 29 and above. In reality, no string of 29 or higher was recorded. Needless to say, the probability prediction was wrong!

Worse, this kind of wrong prediction is not unique to my friend's computer. Just ask any computer science major, he or she will confirm for you that it is indeed true that even the most advanced computer of the distant future will not be able to produce an infinitely long string of random 0's or 1's. So you see, the probability theory fails to predict the absence of long series on all computers.

If you say "What's the big deal? Of course every computer has a limit. No computer can produce endless series. You don't have to tell me that..." then you're already ahead of the field, because many capable men and women -including many capable gamblers of course- have been conditioned by the probability theory to believe otherwise.

Probability problem 3:
It is in conflict with the Law of Large Number

I'm very sure that you have heard of the Law of Large Number. This Law basically states that regardless of short term fluctuations, the long-term ratios will always work out correctly.

It has always been believed that the Law of Large Number is a part of the probability theory; but –as I have shown earlier in the book- this is not true. The Law of Large Number is actually in conflict with probability. And yes, it is a correct law that should be taken very seriously by all gamblers, amateur as well as professional.

The Law of Large Number explains, for example, why in the long run casinos should win. Unfortunately many casino gamblers choose to ignore this important point and end up losing big. A more familiar variation of the Law of Large Number is the Law of Average. Naturally the Law of Average also works against gamblers.

I will return to both of these laws later.

DIFFERENTIATING PROPENSITY AND PROBABILITY

The new concept of propensity

Please note that what I have been attacking is "the probability theory" as taught in high schools and colleges. I'm not attacking the concept of probability, which is the necessary substitute for ignorance when a decision has to be made with insufficient information; as invariably the case in the games of chance. Thus, the concept of probability is alive and well, but the textbook version of probability is partly incomplete and partly incorrect. For this reason, I strongly urge every player in the games of chance -and that means every reader- to un-learn the probability theory and start from scratch with a clean slate and a beginner's mind.

In this new approach to the games of chance we will start with the concept of propensity, then incorporate probability in the second step. As I have discussed in the book, propensity originated from the philosopher Karl Popper as an incomplete concept. However, to pay homage to Popper I have decided to keep the name "propensity" after making necessary modifications to it. What I'm about to present to you is of course the modified and –I believe- more complete meaning of propensity (so that you know what to say just in case some smart alec tries to tell you that you misunderstand Karl Popper.)

The (complete) concept of propensity differs from the textbook meaning of probability mainly in this: It takes into account not only the distributive object (e.g., a coin) but also all other factors (e.g., the process of producing heads or tails) in predicting long-term tendencies, which it identifies as "propensities" (as opposed to the textbook name "probabilities").

We learned earlier that if a robot arm tosses an unbiased coin, or if a person spins a coin instead of tossing it, the long-term ratio for heads may differ from 0.5 by a process-dependent amount. These and more complex variations of distributive processes are naturally accounted for by the concept of propensity, but they would create inaccuracies and require *ad hoc* corrections if analyzed by the methods of the probability theory.

The modified view of probability (variable single-event probability)

With propensity taking over long-term behavior, the role of probability is limited to individual events. Thus, from here on forward, when I say "probability", I mean an educated guess (or prediction if you will) of a single event.

The difference between the new meaning of (event) probability and the textbook probability is this: If we consider a series of apparently equivalent events, the event probability value according to textbooks is the same for all events, whereas the probability value (as modified) may change from event to event.

A descriptive name for this new form of probability is "variable single-event probability". The name can be shortened to "variable probability" without risk of losing its meaning. In fact, I would prefer to call it simply as "probability" when the chance for confusion is zero.

THE ROLE OF LUCK IN A SINGLE BET

The basics of event probability

Do you really believe that the probability is 0.5 for the next coin toss to be heads (as taught by the probability theory). If you do, why bother choosing heads or tails? Why don't just, say, stick with "heads" all the time?

The fact is, we sometimes do think before choosing between heads and tails. This shows that we do not trust the probability theory completely. But what is the reason for this lack of trust? Answer: The failure of the probability to take the total situation into account.

A coin is just an innate object, true, but –as you have seen in the examples of a robotic tossing arm and the process of coin spinning- it would be wrong to assume that, by using an unbiased coin to produce heads and tails in a distributive process, the long-term heads/tails ratio will work out to be 50/50 in all cases.

As we all know, the probability theory considers the coin as the only factor there is. This incompleteness is an example of an unfortunate practice which has been perpetuated over and over in history by a great majority in the scientific community: Factors that cannot be accounted for at the time a hypothesis is formulated are considered as having zero effect on the phenomena under investigation.

In the case of coin tossing this practice leads to the following logic:

1. Major factors in a coin tossing experiment include the coin and the experimental condition.
2. The construction of the coin can be assumed known.
3. The contribution of the experiment is unknown because it varies from case to case.
4. Since (2) can be accounted for, and (3) cannot be; it follows that (3) has no impact on the outcome of the coin tossing process. The analysis therefore should be focused on the coin only.
5. The outcome of the coin tossing process is decided only by the construction of the coin. Moreover, based on past experience with coins of the same construct, the long-term ratio for heads is expected to be 0.5.
6. Conclusion: Since the construction of the coin does not change from toss to toss, and since all other factors have been eliminated, the probability for the next outcome to be heads is 0.5. In fact, the probability for all future outcomes of the same coin is fixed at 0.5.

This is how the probability theory came up with fixed event probabilities. Now that we know that the probability theory is incomplete, let's restart the logic.

1. Major factors in a coin tossing experiment include the coin and the experimental condition.
2. The construction of the coin can be assumed known.
3. The experimental condition is unknown because it varies from case to case.
4. Based on past experience with coins of the same construct, the long-term ratio for heads is expected to be 0.5. It is therefore

logical to assume that 0.5 is the contribution of the coin to the total probability of the toss.

5. The balance of the total probability has to come from the experimental condition.

6. The total probability of the toss is the sum of 0.5 (from the coin) and the contribution of the experimental condition.

Mathematically, if the probability contribution of the experimental condition is positive (leaning toward heads) then the total probability will be in favor of heads, and vice versa. Although in the general case this contribution may not be known, it is expected to be non-zero, simply because an argument for zero contribution would be wrong in many cases.

Not only the probability theory takes the position that factors external to the coin have zero contribution to the total probability of a toss, but also it ridicules the non-zero position as unscientific. However, from the above argument you can see that the unscientific position is that of the probability theory.

In the general case, the contribution of the experimental condition to the total probability is anyone's guess. So if you have a strong urge to choose heads instead of tails in the next coin toss for whatever reason, please feel free to do so because your decision may be as scientific as science can claim to be.

Why the so-called "gambler's fallacy" is not a fallacy, and why it still may not increase your chance of winning

I have just qualified my statement with the phrase "in the general case" because there exist cases where the next outcome in the games of chance can be predicted exactly (100% probability) or almost exactly.

There is a popular strategy adopted by many gamblers. Even gamblers who are well versed with the probability theory sometimes adopts it subconsciously. According to this strategy, while playing roulette one should bet on red when seeing a long string of black, and vice versa. This belief has long been ridiculed by many probability proponents, who gave it the degrading name "The gambler's fallacy".

Why do probability proponents call this popular strategy a fallacy? Because, according to the theory of probability, each spin of the roulette wheel is completely independent of events preceding it. Thus, a long string of black or red is believed to have no consequence on the outcome immediately following its occurrence.

Is the probability theory correct? The textbook answer is "Yes". My answer is a qualified "No!"

Special section: Gambler's wisdom beyond probability

Why "No!"? Because the probability argument is based on the assumption that the process of wheel spinning has no (random) capacity limit. I will argue that this is a wrong assumption.

The roulette wheel is spinned by a dealer. It is the variations in the action of the dealer and the layout of the numbers on the wheel that create the randomness. However, the dealer's motion must appear quite normal without any significant spin-to-spin deviation, unless he or she tries to cheat. It is this "normal action" that forces an element of non-randomness. Thus, the outcome of a roulette spin is decided by a compromise of randomness and non-randomness, not randomness alone.

Are the outcomes of a roulette spin more random than a series of random numbers created by a computer? I'd rather think the reverse is true. After all, the idea of using computer to generate randomness is based on the belief that a computer can be more random than a human being. It is therefore perfectly OK for me to analyze the hypothetical case where black and red are generated by a computer and make the same conclusion for a roulette wheel spinned by a human being.

Now if you take the example of 141.3 billion computer simulated tosses that I cited earlier in the section: "probability problem #2", and assume that bets are made on the choice heads/tails instead of black/red (the minor difference is that there's no provision for the two numbers 0 and 00); whenever you see 28 heads in a row, you can bet all you have on tails without hesitation, because 29 heads simply cannot happen.

In fact you can combine this knowledge with the double-up strategy and start earlier, say, when you see 26 consecutive heads; because you will lose at most 2 and win at least 1 of the next 3 games. I call these cases "pregnant strings". Soon they have to give birth to something else to return to their normal state.

Strictly speaking, then, the so-called "gambler's fallacy" is not a fallacy at all. I would rather call it "the reverse betting strategy". There is nothing fundamentally wrong with this strategy.

But I had to qualify my "No!" answer because reverse betting has almost no practical value in the games of chance; and there are two reasons for this. First, don't forget that although the randomness capacity of every "betting system" is not infinite, it is still huge; so a pregnant string would be a very rare occurrence. Second, it is not likely that the random capacity limit of the "system" that you play against is known to you; so it is almost impossible for you to know if what you're seeing is truly a pregnant string or just a string that appears long to you.

Reverse betting therefore could be a very dangerous strategy. I said "could be" because no strategy is inherently dangerous. It only becomes dangerous when applied incorrectly. Sure, reverse betting could help, and in a very big way too, but only if you happen to be at the right place at the right time. Applying it without knowing its limit is the surest way to get into trouble. My brother told me a casino story that he claimed to be a witness. A woman started playing red exclusively after she saw black turn up 6 times in a row on a roulette table. Black turned up again 13 consecutive times; and before this string ended, the woman had already lost all of her money.

You may disagree with me and reason thus: "I don't need to know the exact length of the pregnant string because a reasonably long string of, say, 6 black implies that the probability for red has increased; and the longer the string of black, the higher the probability for red. Since the normal probability for red is 9/19 (on an American roulette table with 36 numbers plus 0 and 00), with this increase it should becomes better than 50/50; and that's good enough for me."

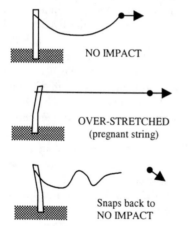

NO IMPACT

OVER-STRETCHED
(pregnant string)

Snaps back to
NO IMPACT

Figure 1s: Distributive process is like the pushing and pulling of a string with one end tied to an elastic pole. Regular pulling motion has no impact (top), but there is a maximum limit that cannot be exceeded (center). When this limit is exceeded, the system snaps back and returns to the unstretched state (bottom).

The reverse betting strategy only works in overstretched situations. But this situation very rarely occurs.

While this "early reverse betting" logic sounds convincing, I'm afraid it is not a valid argument. First, don't forget that the total probability consists of many factor. When you don't have a pregnant string to take advantage of, the probability contribution of other factors may be very significant (I will cover these later.) I suggest that you exercise caution, because this contribution may overwhelm whatever tiny advantage that you may have with early reverse betting.

Second, although I'm attacking the probability theory I must say that, for most practical cases in the games of chance, it is approximately correct. In chapter 3 of this book, I identified the two forces in distributive processes such as the games of chance: Randomness and Symmetry. The probability theory is incomplete because it assumes randomness is the only force at work. The theory survives until today because this assumption is approximately correct for most events in a distributive process. It becomes wrong only when randomness over-extends itself, forcing symmetry to bring it back under control.

I have an analogy to illustrate this point. I compare a series of heads to a force that pulls slowly on a loose string, one end of which tied to a strongly elastic pole. First, the build up of heads has no effect; but when the string start to pull on the pole it will encounter the elastic reaction of the pole, and the situation becomes unstable. Finally, the pulling force is countered balanced by the elastic strength of the pole and heads cannot increase any further. This is where we have a "pregnant string". In the next event, the pole snaps back (by producing a tails to disrupt the long series of heads), and the string returns to its loose position.

In summary, there is no "gambler's fallacy" as alleged by the Probability theory; but reverse betting only works when the system is pregnant with asymmetry and therefore overdue for a reverse. You should have your eyes and ears open to take full advantage if such an extremely rare situation develops; but you should not count on reverse betting as a major part of your winning strategy.

The logical necessity of Luck

When we change money for chips in a casino, invariably before we leave the courteous changer will say to us: "Good luck!" All seasoned gamblers will tell the novice that the best winning strategy is to have Lady Luck on his or her side.

And it's not just in the casinos. We say "Good luck!" to our children before their little league soccer or baseball games, we say "Good luck!" to a friend who is cramming for a bar exam, we say "Good luck!" to another friend who is going for a   job interview, etc... The list is endless.

So does *luck* exist? From the way we deal with life in general and with games of chances in particular, we all seem to agree that it does. In this respect science is very strange. It avoids the subject of *luck* whenever possible, and if confronted it often declares flat out that luck and bad luck are but figments of our imagination.

Why does (existing) science oppose the concept of *luck*? My answer: Because it has been stuck with the probability theory, which equates single event probability with long-term behavior. Since there is only one long-term behavior in a process, each single event is assigned the same probability, leaving no room for *luck*, or for that matter, anything else.

SYMMETRY AND THE END OF PROBABILITY

Since I have refuted the Probability theory and argued that only variable event probability is worth keeping, the next logical step for me is to bring *luck* back to the picture.

Why *luck* and not something else? Because *luck* is a convenient concept encompassing everything that we cannot pinpoint but may contribute to the total probability of an event. It is an attractive concept because, although you and I may disagree on how *luck* comes to existence, we are all familiar with the idea of *luck* and therefore should have no problem agreeing on its impacts.

In simple language, "having good luck" means "doing better than logically expected", "having bad luck" means "doing worse than logically expected". So, in a coin tossing game, although heads and tails turn up more or less equally in the long run, the lucky players somehow either guesses correctly more often than incorrectly, or tends to bet more when guessing correctly and less when guessing incorrectly. Needless to say, the unlucky players tends to do the reverse.

"What if I personally don't think that luck exists?" you may ask. Of course you are entitled to your beliefs and opinions. However, there is a real risk in ignoring *luck* because without it in the equation, you will be tempted to analyze the games of chance by the method of probability and thus deprive yourself the potential opportunity to maximize your wins when you're riding on good luck and minimize your losses when bad luck has decided to accompany you to the casino.

Just remember that *Lady Luck*, in her unpredictable way, may at times be the difference between winning some and losing big. With this power, she commands the respect of all gamblers.

The skeptical gambler may still ask: "If *luck* is such an important factor, why casinos always win?" The answer is: They have the Law of Large Number on their side. I will show that in the next section.

WHY LUCK IS THE BEST WINNING STRATEGY

Luck as a balance between randomness and symmetry

I will admit that I'm not very clear on how Lady Luck operates[1]. Fortunately, I have available to me this tool call "mathematical logic". The great thing about mathematical logic is that it sometimes allows me to analyze things that I know absolutely nothing about. I will now apply mathematical logic to learn more about *luck*.

It would be very naïve for me to think that Nature treats every human being exactly the same way. It would make more sense to think that She treats all of us with the same mechanism or the same logic. In the matter of *luck*, this means Nature will be fair in the totality of her work. However, since each of us is only a part of this totality, we will appear different from one another. As I have argued in the book, the mechanism or logic that can achieve this is a balanced mix of the forces of randomness and symmetry.

Since this has been my main argument against probability, I will not bother you by repeating the details that you can find in chapters 3 and 4 in the book. It suffices to say that each one of us possesses a personal "luck level", and together our luck levels form an approximately normal curve centering around an average value (luck level=0)

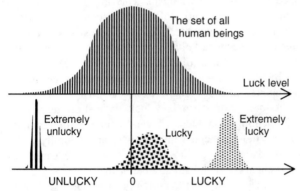

Figure 2s: Nature treats the totality of the human race fairly (top). However since each of us is only a part of the whole, our individual luck varies. A minority of us are extremely lucky (bottom right) while another minority extremely unlucky (bottom left) The rest of us are around the center, with some luckier than others. Note that a reasonably lucky person may have his or her share of unlucky moments, too (bottom center). Just that the lucky moments outweigh the unlucky ones.

As seen in figure 2s, our individual luck level could range anywhere from extremely unlucky (always unlucky when it comes to chances) to extremely lucky (always lucky). In general some of us are more lucky, and some of us more unlucky than the average. Also, it is important to note that a generally lucky person (bottom center of figure 2s) may have unlucky moments; just that these are outweighed by lucky moments. The reverse is true for a generally unlucky person.

A qualification is necessary at this point. The above discussion is completely qualitative and should only be understood as a guideline. For example, it is unclear to me if an extremely unlucky person can exist (on the other hand, I have the strong feeling that an extremely lucky person exists.)

SYMMETRY AND THE END OF PROBABILITY

The reason why casinos don't need luck but gamblers do

Casino profit planning starts with the design of the games. All games are designed to meet a single specification: Giving a reasonable odds to the house.

Mathematically, you can think of casinos simply as luckier-than-average gamblers. The natural question to ask then is: Why casinos always win while other luckier-than-average gamblers may still lose?

There are two reasons for this:

First, thanks to the built-in favorable odds, casinos can afford to have a fixed operational mechanism. This means they don't have to deal with the volatile psychological factors associated with winning and losing, which tend to aggravate long-term losses. I will discuss these factors later in the sections "Five Must-Know Losing Laws" and "Five Must-Know Winning Rules" of gambling.

Second, by playing many many more games than other luckier-than-average gamblers, casinos benefit from the Law of Average, which averages out their short term bad luck with short term good luck, leaving only the long-term odds; which of course are in their favor.

What about those special gamblers who are naturally luckier than the built-in odds of the casinos, or those who are calm enough to successfully enhance their already-better-than-average luck with a "winning system". These people certainly exist, and no doubt at least some of them are attracted to the casinos. And yes, they should beat the casinos even in the long run. That's why once in a while we see in the news stories of fantastic casino wins by extraordinary individual gamblers. Keep in mind, however, that these individuals are very rare. In addition, the casinos are allowed by law to monitor frequent winners and refuse to let them play. What usually happens afterwards is that these winners will end up writing books revealing their "winning system", selling hundred thousands or millions of copies and helping themselves with hefty profits. Most people who buy these books go on losing because they forget that the writers are either very lucky or very disciplined in addition to being lucky enough. The average gamblers simply do not have these qualifications.

You should also keep in mind that most of the very lucky people don't hang around casinos because they are too busy making big money elsewhere. This and the casinos' efforts to push the few potential winners away lead to the result that, the average casino gambler is less lucky than the average person in the society at large.

It is no surprise then, that casinos always win!

By symmetrical logic, you can immediately see two disadvantages for the gamblers:

First, the built-in unfavorable odds force gamblers to make subjective decisions. Although these subjective decisions may be based on clever schemes that actually reduce the casino's advantage, the problem is, human beings are very emotional by nature. The casinos of course are designed with an atmosphere to make sure that this emotional aspect is exaggerated to the max. As a result, tactical and strategic errors are constantly committed, even by those have vowed not to commit them ever again. I will discuss this in more detail in the section "Five Must-Know Losing Laws of Gambling".

Second, because of the unfavorable built-in odds, the average gamblers are at the short end of the Law of Average. The more games he plays, the higher the probability that he will lose.

But this makes the short run interesting because here the Law of Average does not apply. The highly unpredictable nature of the short run makes luck the overriding factor, and the average gambler may in fact win. I will return to this important point later in the section "Five Must-Know Winning Rules of Gambling".

The logic of winning and losing in the games of chance

Make no mistake about it. Casinos always win because –most of the time and certainly in the long run- the average gambler would lose to them.

But who is this "average gambler"? There is an argument that, because of the Law of Average, over time every gambler is an average gambler; and therefore every gambler will lose to the casinos. I hope that my analysis of individual luck has shown clearly that this argument is a fallacy.

Let's me give you a personal account before I continue with the discussion. I personally know a gentleman who would go to Reno once every week for over twenty years now, and his net adds up to a huge sum. It is therefore important to realize that, although most gamblers are losers, a minority of them do win.

It is quite possible that you belong to this small circle of winners; but even if you're not, knowing the logic of winning and losing in casino games should help you cut your losses and make your trips to the casinos more enjoyable.

For this purpose I will devote the rest of this chapter on the "Five Must-Know Losing Laws" and the "Five Must-Know Winning Rules" of gambling. I recommend that you read them carefully and think about them seriously before making your decisions.

I will break tradition and present the losing laws first. While this appear to be a pessimistic method of presentation, there is a good reason. Gambling is like a military expedition. Usually the side that understands the mechanism of losing and is more afraid of losing will emerge victorious.

FIVE MUST-KNOW LOSING LAWS OF GAMBLING

The Law of Average
and why beginner's luck could be a very dangerous thing

I have already discussed the Law of Average at great length. I would like to emphasize the implications of this law in the special case of the first-time gamblers. There is a strange phenomenon known as "beginner's luck"; and based on my own observations for over more than 30 years, this is not just a myth. On the average, Lady Luck seems to treats beginners better than the experienced gamblers.

Some addicted gamblers who were once lucky beginners theorize that beginner's luck is the work of the Devil. How could such a great thing be associated with such a horrible symbol? That is because many lucky beginners do not realize that they are lucky and actually convince themselves that they are born winners in the games of chance.

A minority of these people may be right, but for the majority this is a one way ticket to disaster. Over time most lucky beginners will lose like everyone else, because the Law of Average will make sure that their initial luck is compensated for later on by bad luck. The problem is, the great first day has left such an indelible impression in these gamblers' mind that it is almost impossible for them to admit that they will continue to lose.

The Law of Decreasing Value
and why gamblers tend to increase their bets with time

When I was an undergraduate, one day my roommate and I figured out that we would not have enough money to pay the rent for the next several months. It so happened that there were several wealthy students living close by who were interested in playing stud poker for money. It was during this short period of time that I learned a great gambling lesson from my roommate. To cut a long story short: I made some money thanks to this lesson.

The lesson is about the decreasing value of money in a long gambling session. Like all other secrets, the truth contained in this lesson is very obvious once you know it. In fact you can confirm it either by observing your own behavior or silently watching other gamblers in a casino over a period of several hours.

More often that not, a typical gambler would start out with a relatively conservative betting strategy; and then as time progresses he would becomes more and more aggressive, betting more and more money per game (provided that he still has enough money).

What about those gamblers who suffer big losses because of their increasing aggressiveness? A minority would become even more aggressive, and only back down if they are lucky enough to win back a substantial percentage of their losses. But the majority is more interesting. They would temporarily reduce their bets, and regardless whether they will win or lose the next series of games, they would at some point increase their bets again.

I call this phenomenon "the Law of Decreasing Value". It can be stated as follows:

"To the typical gambler's mind, a given amount of money appears to decrease in value over time in a long enough gambling session."

The Law of Decreasing Value is an extremely important law. For example, it helps the uncommitted observer to understand why the same gambler who is hesitant to bet $15 at the beginning of his blackjack session suddenly raises his bet to $30 after say 50 games. More importantly, I hope it helps you to understand your own psychology on the fly in a gambling environment and make adjustments as necessary.

Why do you need to analyze your own psychology? Because the Law of Decreasing Value spares no one, except the uncommitted gamblers, and I haven't seen any uncommitted gambler around a blackjack table for a long time.

The importance of the Law of Decreasing Value can never be emphasized enough. Not only it usually makes losers lose much more money than they should, but also it can quickly turn winners into losers. The mechanism for the first case is obvious. I will therefore discuss only the second.

Suppose after one hour a typical gambler (i.e. one not blessed with extraordinary luck) has won about US$500 by averaging $30 per bet. If he keeps fluctuating around this amount, probably everything will end well. But what if the dealer goes on a streak, and after a few unsuccessful splitting and doubling efforts the gambler suddenly sees that his winning has dwindled to $250? Here is the dangerous moment where the Law of Decreasing Value comes into play. Most likely the gambler will raise his average bet to, say, $50 without realizing that this action plays right into the hands of the Law of Average, which of course favors the dealer.

Why does the Law of Average favors the dealer at this particular point? Because higher bets means it take fewer games for the gambler to lose the amount that he has won with so many games earlier. And since he has won so many games earlier, by the Law of Average he is overdue for a reverse of fortune. Yes, he can be saved if Lady Luck is still on his side; but She is rarely on the side of the average gambler after he has won many games.

Unless it is an extremely lucky day for the gambler who won early, the Law of Decreasing Value and the Law of Large Number usually gang up to make him a loser at the end.

The Law of Vanishing Value
and why gamblers tend to overstretch their luck

Let's suppose for argument's sake that Lady Luck decides to stay; and by raising his bets, the gambler gets back to the $500 that he won originally. Does the danger end? Hardly! Now he feels lucky. He may even feel a sense of superiority over the dealer. He most likely will stay with the high bets. He may even increase it, risking losing all he has won, and more. This vicious circle will go on and on until the end of the session.

But because this is a hypothetical case, let's assume that, somehow all the chips fall in place, and the high flying gambler quickly wins a substantial additional sum, say $10,000 (for a total winning of ten thousand five hundred.) Now what happens next?

There are two possibilities. If the gambler somehow realizes that he has been very lucky and quits when he is still ahead, everything would be fine. But this is not how the typical gambler's mind operates. To him, he has just won $10,000 more than expected and therefore could afford to lose up to this amount. In other words, the gambler considers the sum of $10,000 that he won in his wild streak as *free money*. Having free money and riding on luck, why not try and see what happens? "The worst than can happen is that I will lose all of the free money, so what! It wasn't mine to start with anyway." Thinking thus, the gambler will stretch his luck; believing this cannot hurt.

As you can see, there is no such thing as a "long streak of luck" for the average gambler because no matter how long it is, he will stretch it further until it snaps back on his face. Eventually his fortune will reverse. It must, because this is what the Law of Average is all about.

I call this phenomenon "The Law of Vanishing Value". It can be stated as follows:

"The value of money drops to almost zero and stays there until the big winner loses all of the money that he won."

Even when the big winner has lost all of his winnings, he is still not off the hook, because the Law of Decreasing Value is still in effect and will affect his betting strategy with his very own money.

Usually the Law of Vanishing Value only affects gamblers who have just won big; but "winning big" is a subjective notion. It is therefore prudent to examine your own psychology to make sure that you're not afflicted by this Law.

The Law of Symmetry
and why most winning gamblers will eventually lose

Some gamblers keep winning for a long time. I personally know a few who even bought houses with their casino and stock winnings. Except for the very small minority who decided to leave gambling permanently (usually thanks to the positive influence by their loved ones), I notice that all who enjoy spectacular gambling successes have one thing in common: They gamble with higher and higher stakes!

One time in the 1990's when I had just returned from Asia after a long business trip, I was told that a friend of mine had just won more than US$100,000 in stocks the other day by playing options. Before I had the time to call and congratulate him, I was told that he had just lost about $200,000 in a speculation completely outside his field of knowledge. This meant he lost not only the $100,000 that he had just won, but also an additional $100,000 from his own "war chest". The guy was a well to do businessman, but he was by no means a multi-millionaire. It was truly amazing to think that someone would place an amount of the same order of magnitude as his total net worth in a single bet on something he does not know enough about. But that is what long-time gamblers always seem to end up doing. They act as if they try to figure out a way to lose.

Since these special high rollers are usually very smart people and are fully aware of the risk they are taking, this is not a result of either the Law of Decreasing Value or the Law of Zero Value. So what is it? My answer: The Law of Symmetry!

One great prediction of the Law of Symmetry is that every asymmetrical situation has a mirror image, and if the process develops long enough, the two will average each other out. Amazingly, this is also true in gambling. When you think about it, winning big over a long period of time is a highly asymmetrical situation. If the winners kept making the smart decisions that led to their winning record, how could this highly asymmetric situation be corrected? The mechanism for the Law of Symmetry in gambling therefore is on the edge of a mystical law. It basically says that dumb mistakes will have to be made by smart people in the long run so that symmetry is obeyed. I will not speculate more on it. It suffices to just state this law as follows:

"Most long-term winners in the games of chance will behave in such a way that the probability of losing everything they ever won is maximized."

SYMMETRY AND THE END OF PROBABILITY

Under the spell of the Law of Symmetry, even the long-time winners usually lose at the end.

The Law of Increasing Randomness
and why almost everyone will lose to "the system"

The four laws cited can be combined into a single law "The Law of Increasing Randomness". Here increasing randomness refers to the behavior of the gamblers over time. The Law of Increasing Randomness can be stated as follows:

"Regardless whether a typical gambler is winning or losing, his gambling behavior will become more and more random over time. Consequently, his chance of losing becomes more and more certain over time."

The Law of Increasing Randomness is a most interesting law because it is the gambling parallel of the Law of Entropy of thermodynamics. In fact it can be shown to be a social equivalent of the Law of Entropy; but since a proof is not trivial[2] I have decided not to burden the reader with it.

In a nutshell, the Law of Randomness says that most gamblers are programmed to lose when they play against a system (e.g. the casino). The case-to-case difference is just a matter of time.

Does this mean every average gambler will lose? Not necessarily. I will explain why in the next section "Five Must-Know Winning Rules of Gambling".

FIVE MUST-KNOW WINNING RULES OF GAMBLING

Gambler's winning rule #1:
Answering the question "Am I fit to gamble?"

In case you have forgotten what propensity means, it is simply the long-run tendency of a distributive process. In the case of casino gambling the process is "the individual versus the casino". The individual is of course you. Since we are dealing with gambling, I will focus on "gambling propensity".

Why "gambling propensity" and not just "propensity"? Because a generally lucky person (i.e. one who has good general propensity) may turn out to be very unlucky in the specific type of gambling that he or she may get into (i.e. one who has bad "gambling propensity"). That is because there is a big difference between "general propensity" and "gambling propensity": Winning and losing is only a part of life, but winning and losing is everything in gambling!

Take a successful businessman who is happily married and blessed with beautiful children for example. We must say that his general propensity is good to excellent. But don't forget that in the calculation of this "general propensity", gambling luck has not been taken into account. It would be a fallacy to assume that since a person is generally lucky in life, he should also be lucky in gambling. A more logical analysis must be based on past situations that bear some resemblance to gambling: They are situations where the outcomes are not clear ahead of time, and there are winners and losers at the end.

Just from this simple analysis, you can judge for yourself if you may fit in with gambling. If you have lost more often than won in social games of chances, my advice to you regarding gambling is "Don't start"; because starting gambling means that you will take on a formidable adversary with an excellent track record and all the odds on its side (i.e. the casino). It would be ridiculous to assume that a person with a losing record could beat such an adversary, especially in the long run.

On the other hand, if your winning/losing record is average or above, you should have a fighting chance against the casinos. Who knows, you may even be lucky enough to beat the house in the long-run. But even if this is the case, I still strongly recommend you to focus on rules #2 and #3 before fixing your eyes on gambling.

Gambler's winning rule #2:
Deprogramming the 5 losing laws of gambling

If you have skipped the Five Must-Know Losing Laws of Gambling I strongly recommend you to read them carefully, think about them seriously, and consider them as time tested cautions or warnings. I know that we all have the tendency to say: "But the mistakes that you talked about are so stupid. A smart person like me would never commit these silly mistakes!" To this objection I would like to remind you that history is filled with extremely stupid mistakes made by extremely intelligent men and women; and history does repeat itself.

The Five Losing Laws of Gambling point to one major point: The average gambler is programmed to lose! Can a serious gambler do anything about this terrible fate? My answer is "Yes!" and the key word of my answer is: "Deprogramming"!

"Deprogramming" is not just a buzzword, it actually is a meaningful action that can be achieved. Think of the 5 Losing Laws as terrible virus programs that are secretly built in as a part of the "universal subconscious" common to all human beings. There is nothing carved in stone that says we cannot fight back to at least keep these viruses from being activated. To fight a virus effectively, the first step is to analyze its characteristics and how it may cause harms. I have done so earlier for the 5 Losing Laws. The second step is to use this information to nullify or minimize its effect. I don't believe nullification is a practical goal. However, I believe that everyone of us has the ability to minimize the effects of the 5 Losing Laws, provided that we have studied them to the point that we can say we have understood their mechanisms.

The main losing mechanism in all 5 Losing Laws is the increase in randomness in the gambler's behavior. While it may be impossible to get rid of this tendency completely, it is possible to train yourself to watch out for danger and set up strict personal guidelines ahead of time. This is easier said than done; but it must be done before any serious gambling should be attempted.

I want to end this section with a bittersweet true story. A very savvy day trader was able to achieve something that most other traders with middle class income in the US could only dream of in an economic downturn. He won $300,000 in a short time by betting big on put options. Do you see anything here? The guy was following the exact pattern programmed in the Law of Increasing Randomness because, with a middle class income, he must have betted too much in the market in order to win that much. Guess what? He ended up losing all he had won plus $100,000 of his own money also in a short time. Fortunately, he got married. Family life changed him completely. First, he quickly recovered psychologically from the financial disaster he had got himself into. Second, he learned his lesson and decided to stick to the conservative rule of never betting more than $500 on any day.

Let's hope that he continues exercising constraints and keeps having a happy family.

Gambler's winning rule #3:
Choosing and learning a strategy that fits your personality

Although casino games are designed to give the odds to the house, with proper strategy and practice, these odds can be reduced to the point that, with a little luck, a gambler could beat the house.

There are all sorts of books on gambling strategies. There is at least some truth in all of these books, because they are usually written by exceptional gamblers. However, the very fact that the writers are exceptional gamblers –and it would serve you well to make the modest assumption that you're not- means that it is very difficult to apply their strategies as intended.

One way to reduce this difficulty is to choose a strategy that fits your personality instead of a strategy that sounds most attractive. "Why my personality and not my ability?" you may ask. My answer is: Because we're very emotional beings and, like I have said earlier, the casino environment is designed to exaggerate our emotions to the max. Since fortunes could be made and lost at a wink of an eye on a casino table, it is usually habit, not intellect or ability, that dictates your choices in critical moments.

By studying and practicing the appropriate strategies you will be able to narrow the gap with the casinos. Mathematically they may still seem to have a long-term advantage, but don't forget that you have several options that are not available to them: You can choose when to play and when to quit, you can choose to raise or lower your bet, and (if your sixth sense is good) you may be able to synchronize your options with the ever unpredictable behavior of Lady Luck and bring Her to your side.

If managed properly, these options –which will be covered in rules #4 and #5- should help minimize your losses; and if Lady Luck is reasonable, you may even be able to wipe out whatever advantages the casino still has over you and emerge a winner.

Gambler's winning rule #4
Combining short sessions with fixed winning and losing limits

In gambling, a display of ego is a sure sign of imminent defeat. Even the most confident gambler should exercise modesty and plans as if he is an average player. This way he won't get caught by surprise and suffer big losses.

Earlier I have argued that the best chance for the average player to win is in the short run. It is time for me to be more specific. Figure 3s shows three major situation: The average gambler certainly will lose in the long run and most likely will lose in the medium run. This leaves only the short run, where there are 3 situations, and the average gambler does win in one and break even in one. From this analysis, it is clear that the optimum gambling strategy is to plan for the short run.

But what is a short run? From my analysis in chapter 5 of this book, a short run is 100 trials or less. This would mean 100 blackjack hands or less, for example.

You may ask "How can I plan the short run when I spend my whole weekend at the casino?" The easy answer is "Then don't spend the whole weekend!" However, I do realize that this is not a reasonable advice. The other alternative is to plan other activities such as sight seeing, going to musical shows, observing and learn from savvy players, etc. Gambling doesn't have to be the only fun thing to do at the casinos.

But if you really want to play more than 100 hands, there is a workaround: Take a twenty minutes break after completing a short run, then restart. While this doesn't sound like a spectacular idea, it actually gives you two great advantages that very few gamblers are aware of:

1. It interrupts your tendency to fall in the trap of the Law of Decreasing Value, which favors the dealer.
2. It interrupts the action of the Law of Average, which also favors the dealer.

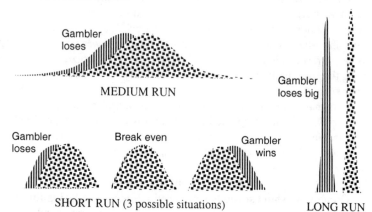

Figure 3s: In these pictures the winning edge belongs to the curve on the right, with vertical lines representing the average gambler and dots the casino. In the long run and medium run it is clear that the house will beat the average gambler. The only chance for the average gambler is the short run, where he may be even with the house. The gambler's first trick is to convert the long run to a series of short runs to nullify the house's advantage. This can be done but only if the gambler can deprogram the effects of the 5 losing laws, which are in the subconscious of every gambler.

Thanks to these little known advantages, short breaks help to simulate short runs. Adding the betting skills that you have learned in a system of choice (mentioned in rule #3) to the short run strategy, you may be able to break even with the house; meaning that your winning trips and losing trips are approximately equal in number over time.

Now comes the most rewarding part in the plan. The mathematics for this part is simple. Given that the number of winning trips and losing trips are approximately equal, if the average win on a winning trip is $W and the average loss on a losing trip is $L, the average winning per trip is ($W-$L)/2.

All you have to do now is to work backward. For example if you want to average $200 win per trip, $W will have to be larger than $L by $200×2=$400. One set of numbers that satisfies this condition is $W=$1200 and $L=$800. If these numbers fit your budget, the last step is to set the following condition to yourself:

1. Stop playing if the total win in a trip reaches $1,200.
2. Stop playing if the total loss in a trip reaches $800.

Sounds very elementary, doesn't it? But that's exactly why it works like a charm. I know only a few people who win consistently over time. And short run with clear limits is exactly their winning secret.

But like every other success formula, keeping the same limits consistently is a very challenging task. You will have to exercise tremendous discipline. But if you're serious about gambling, the reward of winning should be a strong enough motivation to help you achieve this goal.

Gambler's winning rule #5
Knowing a little more about Lady Luck

I will not go into betting strategy for individual games because I assume you have learned enough from selected books on gambling tips and already settled on a system fitting your own personality. But since luck is the dominant factor in a gambler's success, I think it is necessary to clear up a few things about Lady Luck.

First, I will not claim that I know anything about Lady Luck's behavior because I'm myself at Her mercy whenever I go to the casino. However, from the scientific concept of propensity as a total gambling condition, I was able to deduce a few dos and don'ts that I want to share with you.

Remember that the only propensity that you're interested in when you gamble is the difference in propensities between you and the casino (or the dealer, who represents the casino.) If this difference is in your favor, you will win even if everyone around you are losing; and if it is in favor of the casino, you will lose even if everyone around you are winning.

Although the value of this difference at any given time is completely unknown, by the law of cause and effect I will argue that it may switch side (from positive to negative or vice versa) whenever there is a noticeable change in the overall condition. In other words, any noticeable change in the overall condition of a gambling table may change a winner to a loser, and vice versa. By "noticeable change" I meant any event that catches your attention.

If you have a scientific background, you probably will say that this cause-and-effect reasoning is just a nonsense excuse for superstition. But I would like to remind you that we are dealing with unknowns. When you have to deal with an unknown situation, the only method that has any practical value at all is the engineering method, which boils down to the procedure of trials-and-errors on all events that you feel may be significant, regardless how insignificant or frivolous they may appear to be. After all, since luck is so important and there is no science that can teach you the cause-and-effect of luck, why not go with your feeling and intuition? You have to make your own decision here; but I can tell you that if I have to choose between being called a "superstitious winner" and a "scientific loser", I will choose the first without blinking an eye.

Just think about this. Casinos are owned and managed by the most practical people in the world. Then why it is a common casino practice to change the dealer when the house keeps losing on a blackjack table? And why most of the time this strategy works? Scientists call this kind of practice "rubbish"; but engineers are much more practical; they simply accept reality. Since the goal of a gambler is extremely practical, I advise every gambler to follow the practice of good engineers, especially because the casinos have done so already.

Once it is agreed that a change in overall condition may result in a switch in luck, the casino strategy of changing dealers after a long losing streaks that I've just mentioned becomes self explanatory. The practical lesson for the gambler in this case is: Reduce your bet at least temporarily (or change to another table) when the casino switch dealer after they've suffered a long losing streak.

Many players employ the following strategy: They look for a table where the dealer is losing and jump in, assuming that by doing so they can ride on the house's bad luck. In my opinion this strategy is ill-founded. First, when you join a table, your own participation may change the total condition and switch the house's luck. Second, a dealer who loses to others may not lose to you.

The best strategy is to open your eyes and ears when you play. Hopefully over time you will be able to identify a table that you tend to do well, dealers who tend to break even or lose to you. And you may even develop a sense for when there is a change in condition, such as a player leaving the table, new player joining in, changing of dealer, etc.

I personally believe that Lady Luck is not only unpredictable, but also very selective. She only helps those who respect and spend their efforts trying to understand Her.

GOOD LUCK!

NOTES:

1. There are fields that claims to have answers to these questions, but their status are now at best pseudo-sciences.

2. Strictly speaking, the Law of Entropy applies to closed systems (no in or out) under equilibrium. These conditions don't seem to be met by the environment where gambling takes place (e.g., casino + gamblers). That's why establishing the parallel between the Law of Increasing Randomness and the Law of Entropy is not trivial.

References

"Probability, Statistics, and Truth" (book), Richard von Mises, Dover 1981. This is an unabridged reproduction of the 1957 English translation of the 1951 German original; which in turn is a revision of the 1928 German original.

"The Logic of Scientific Discovery" (book), Karl Popper, German original 1935, English translation Harper Torchbooks, 1965

"Propensity, Probabilities, and the Quantum Theory" (article), Karl Popper, 1957; reprinted in "Popper Selections", edited by Miller, Princeton University Press, 1985.

"Religion and Science" (book), Bertrand Russell, Oxford University Press, 1997 (first published by the Home University Library, 1935)

"A Physicalist's Interpretation of Probability", Laszlo E. Szabo, talk presented at the Philosophy of Science Seminar, Eötvös, Budapest, 8 October 2001.

"Sync – the Emerging Science of Spontaneous Order" (book), Steven Strogatz, 2003, THEIA (Hyperion books).

"Synchronicity: An acausal connecting principle" (book), C.G. Jung, English translation by R.F.C. Hull, third edition, Princeton, 1973.

APPENDIX: ALL THE MATH YOU NEED TO KNOW

It is assumed that you know how to add, subtract, multiply, and divide two numbers. It will help if you know what a "linear equation" is (but if you don't, you may still get away with it.)

<u>GENERAL RULE:</u> Some equations are included in the book for scientific rigor. You should glance at them, but you don't need to understand them. Just make sure that you read the texts that explain their meanings. You will find diagrams and drawings helpful. By all means don't skip them.

<u>Dimension</u>: A point has zero dimension, a line has one dimension, a surface has two dimensions, a volume has three dimensions.

<u>Positive and negative numbers:</u> A convention to convey the idea of opposites. Instead of saying A is 5 meters from my right and B 5 meters from my left, I'll say A=+5 meters and B=-5 meters to convey the same idea (A=-5, B=+5 are OK, too.)

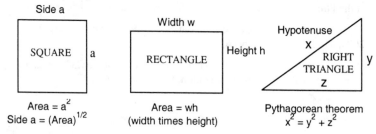

Side a

SQUARE a

Area = a^2
Side a = $(Area)^{1/2}$

Width w

RECTANGLE Height h

Area = wh
(width times height)

Hypotenuse
x
RIGHT TRIANGLE y
z

Pythagorean theorem
$x^2 = y^2 + z^2$

<u>Power:</u> a^n is called "a to the power n" or "a to the nth power". It is *a* multiplied by *a*, then again multiplied by *a*, etc. so that *a* occurs *n* times. Example: $5^3 = 5 \times 5 \times 5 = 125$.

Important:

1. If $a^n = b$, then $a = b^{1/n}$. Also "a squared" means a^2 = a times a, "square root of a" means $a^{1/2}$. Example: 5 squared = 5^2 = 5x5 = 25, square root of 25 = $25^{1/2}$ = 5.

2. $[a^n]^m = a^{nm}$. Example: $[5^2]^{1/2} = 5^{(2)(1/2)} = 5^1 = 5$.

The idea is extended to cases where n is not an integer (not a whole number). Thus, quantities such as $5^{2.3}$, $7^{1/0.35}$ are permissible (you will need a calculator to figure out the values.)

<u>Logarithm (lnx) and exponent (e^x):</u> These are two "functions" of x; meaning that for a given value of x there is a unique value for lnx and e^x. Don't worry if someone ask you "what is log of 5" and you don't know the answer. Nobody else does. That's what calculators are for.

<u>Plus minus sign ($\pm$):</u> If x = a $\pm$ b then the value for x is between (a-b) and (a+b). Example: x=5 $\pm$ 2 means x is between 3 (because 3=5-2) and 7 (because 7=5+2).

<u>Fourier transform:</u> This is a mathematical operation. You don't need to know what it is.

SECOND BOOK IN THE COMMON-SENSE SERIES
To be published in 2003

The Symmetry Foundation of Quantum Physics

DangSon Tran

Quantum mechanics, arguably the most glorified branch of science, is full of paradoxes and crazy consequences: Particles do not exist until we measure them, particles exist only on borrowed time, a cat trapped in a cage with a vial of poison hooked to a radioactive material will be *both dead and alive*, our world splits into multiple worlds each time a choice has to be made, and many more other scenarios that do not make any sense at all.

Einstein never accepted quantum mechanics as a complete science, but he could not disprove it. Many quantum mechanicists felt something was seriously wrong with their science but could not figure out what it was.

It never occurred to these experts that the root cause of all "quantum weirdness" has to lie at the foundation of quantum mechanics, namely the Probability theory. This simple realization was the missing key that led to the remarkable breakthroughs reported in this book; breakthroughs that will change science forever.

By solving all major problems, puzzles, and paradoxes associated with quantum mechanics, this book is a conclusive proof that it is human insight, and not complex mathematics, that holds the key to the future of science. While some mathematics was necessary for scientific rigor, the book was written in a layman style accessible to everyone who has an inquiring mind.

This book will start a new scientific revolution and you –the reader- could be an active participant in it. All you have to do is to open your mind and listen to the little child of common sense inside your own self.

The Science of Space-Time and Existence

DangSon Tran

Einstein's celebrated theory of Relativity also has its share of paradoxes. Best known is the "Twin paradox". According to Special Relativity, a moving clock will run more slowly than a stationary clock. Suppose one of the twins decides to take a long trip in a fast moving spaceship, which one of the twins will age more slowly? The quick answer is "the traveling brother".

The sticky point in the Twin paradox is the Relativity position that all motions are relative. If the stationary twin sees his brother moving at velocity v, the moving twin will see his brother moving at velocity –v. Since the ticking rate of clocks only depends on v^2, it appears unclear which one of the twins will be younger over time.

Some experts say that the situations of the twins are not equivalent. They argue that the traveling brother has to accelerate when he leaves home, decelerate when he is ready to return, accelerate to return, then decelerate to stop at home. These accelerations and decelerations make the motions of the two brothers inequivalent. This is claimed to be the key to solve the paradox.

The problem is, in addition to avoiding the simple but troublesome symmetrical condition of the problem, this solution requires the existence of the rest of the universe. What if all is left of the universe are the two spaceships? It turns out that there is no satisfactory answer in the theory of General Relativity.

The fact that all observers, regardless of their state of motion, see the same (constant) speed of light is also serious paradox, although scientists prefer to call it a mystery. What is so special about light that allows it to dictate length scale and time scale in classical physics? No one, including Einstein, ever offered an answer to this question.

Entropy is another controversial subject. The Distribution theory has resolved one issue. Entropy is a manifestation of deterministic distribution, and has nothing to do with probability! It should follow that classical time will always flow forward; but isn't this in conflict with Einstein's concept of a space-time continuum that apparently allows time travel?

At the center of all these issues is the question of space and time. It appears from the first two books of the "common sense" series that space and time have very limited role in quantum processes; but that is because we have not touched the subject of existence. The fact is, existence –as we know it- cannot be defined without reference to space and time; and that is true for everything, from the micro-world of atomic particles to the vast expanse of the universe. Einstein asked us to relinquish common sense and accept his idea of a space-time continuum. The problem is, among other things, Einstein's space-time continuum does not seem to be a faithful description of the microscopic realm of particle physics.

Could it be that Einstein gave up on common sense too early and therefore missed something fundamental? While the pundits say it is impossible, we will seek answers to all of the above and many other exciting questions in "The Science of Space-Time and Existence". How will we do it? You guessed it right! With common sense!

INDEX

Q

R

S

T